Electrochemical Deposition: Properties and Applications

Electrochemical Deposition: Properties and Applications

Editor

Gergana Alexieva

MDPI

Basel • Beijing • Wuhan • Barcelona • Belgrade • Novi Sad • Cluj • Manchester

Editor
Gergana Alexieva
Sofia University "St. Kliment Ohridski"
Sofia
Bulgaria

Editorial Office
MDPI
St. Alban-Anlage 66
4052 Basel, Switzerland

This is a reprint of articles from the Special Issue published online in the open access journal *Coatings* (ISSN 2079-6412) (available at: https://www.mdpi.com/journal/coatings/special_issues/Electrochemical_Properties).

For citation purposes, cite each article independently as indicated on the article page online and as indicated below:

Lastname, A.A.; Lastname, B.B. Article Title. *Journal Name* **Year**, *Volume Number*, Page Range.

ISBN 978-3-7258-0193-0 (Hbk)
ISBN 978-3-7258-0194-7 (PDF)
doi.org/10.3390/books978-3-7258-0194-7

Contents

coatings

MDPI

Article

Structural, Morphological and Optical Properties of Nanostructured ZrO_2 Films Obtained by an Electrochemical Process at Different Deposition Temperatures

Konstantin Lovchinov [1], Rositsa Gergova [2] and Gergana Alexieva [3,*]

[1] Acad. J. Malinowski Institute of Optical Materials and Technologies, Bulgarian Academy of Sciences, Acad. G. Bonchev Str., Bld. 109, 1113 Sofia, Bulgaria; lovchinov@iomt.bas.bg
[2] Central Laboratory of Solar Energy and New Energy Sources, Bulgarian Academy of Sciences, 72 Tsarigradsko Chaussee, 1784 Sofia, Bulgaria; rgergova@phys.bas.bg
[3] Faculty of Physics, Department of General Physics, University of Sofia, 5 James Bourchier Blvd., 1164 Sofia, Bulgaria
* Correspondence: gerry@phys.uni-sofia.bg

Abstract: This article focuses on the impact of the deposition temperature (in the range from 60 to 80 °C) in ZrO_2 films obtained by the electrochemical deposition process on SnO_2-covered glass substrates. The solution in which the deposition takes place is aqueous, containing $ZrOCl_2$ with a concentration of 3×10^{-5} M and KCl with a concentration of 0.1 M. By implementing X-ray diffraction (XRD), optical profilometry, scanning electron microscopy (SEM), and UV-VIS-NIR spectroscopy, the temperature dependence of ZrO_2 films properties was revealed. The X-ray Diffraction XRD spectra showed six different diffraction maxima ((−111)M, (101)T, (111)M, (112)M, (202)M, and (103)M) associated with the electrochemical ZrO_2 layers, and the polycrystalline structure of the films was confirmed at all deposition temperatures. The determination of the average roughness did not indicate significant temperature dependence in the deposited layers. SEM micrographs showed that the layers were composed of grains, most of them of a regular shape, although their size increased slightly with an increased deposition temperature. The coarsest-grained structure was observed for the layers deposited at 80 °C. It was demonstrated that the deposition temperature weakly impacts the reflectance and transmittance spectra of the ZrO_2 layers. Such layers with low values of specular and high values of diffuse transition, and reflection in the spectral range from 380 to 800 nm, can be applied to various optoelectronic devices such as thin-film solar cells.

Keywords: electrochemical deposition; surface morphology; ZrO_2 films; optical measurement

Citation: Lovchinov, K.; Gergova, R.; Alexieva, G. Structural, Morphological and Optical Properties of Nanostructured ZrO_2 Films Obtained by an Electrochemical Process at Different Deposition Temperatures. *Coatings* **2022**, *12*, 972. https://doi.org/10.3390/coatings12070972

Academic Editor: Saulius Kaciulis

Received: 8 June 2022
Accepted: 4 July 2022
Published: 8 July 2022

1. Introduction

Metal oxide nanostructured materials are traditional but increasingly relevant subject of many scientific and technological researches. Recently, zirconium dioxide (ZrO_2) nanomaterials have attracted considerable scientific interest due to their inherent combination of properties such as high mechanical and thermal resistances, chemical inertness, high corrosion resistance, high refractive index, wide range of optical transparency (UV–Vis–IR region), and good dielectric constant [1–7]. Also known as "zirconia," zirconium dioxide is a wide-bandgap n-type metal oxide semiconductor with high ionic conductivity [8–10]. ZrO_2 is a ceramic material. At atmospheric pressure, it occurs in three phases depending on the formation temperature: monoclinic (below 1170 °C), tetragonal (between 1170 °C and 2370 °C), and cubic (above 2370 °C) [11].

Because of its unique combination of electronic and mechanical properties, ZrO_2 proves to be a promising material for use in many applications including gas sensors [12,13], solar cells [14], dental implant materials [15], fuel cells [16,17], etc.

A considerable number of research articles report ZrO_2 structures obtained by different methods, such as the following: sol–gel method [18], magnetron sputtering [19,20], chemical vapor deposition [21], electron beam physical vapor deposition (EBPVD) [22], laser ablation [23], hydrothermal method [24], and atomic layer deposition [25]. A key element in the design of zirconia nanostructures is the choice of technique and approach for their elaboration, which ensures the presence of needed properties and qualities, and predetermines specific applications. Therefore, in this research study, the emphasis was placed on the electrochemical technique and the influence of the deposition temperature (in the range from 50 to 80 °C) on properties of ZrO_2 thin films deposited on SnO_2-covered glass substrates. The electrochemical method can be applied to produce large areas with controlled and diverse morphology and size nanostructures. This method is quite inexpensive, and it does not require the use of complicated equipment.

In the presented manuscript, we report the results for structural, morphological, and optical properties of the nanostructured ZrO_2 films under investigation, amplifying the information already existing in the literature for similar layers produced by electrochemical deposition technique [26,27].

2. Materials and Methods

This section describes the method of deposition and the conditions under which it was carried out, as well as the specific equipment that was used to obtain the results of the research. The deposition of the zirconia layers was electrochemical and was carried out by means of three electrode cells located in a thermostatic bath (Figure 1). The deposition took place in an aqueous solution. Dissolved in this solution were $ZrOCl_2$, supplied by Alfa Aesar (98%, with concentration of 5 mM) and KCl supplied by Valerus Ltd. (99.5%, with concentration of 100 mM). Deposition was realized at different electrolyte temperatures in the range from 60 to 80 °C, and the potential between saturated calomel electrode (SCE) and substrate (cathode) was kept constant at -700mV. The sample in the solution is in thermodynamic equilibrium. The deposition time of all zirconia layers studied was 20 min.

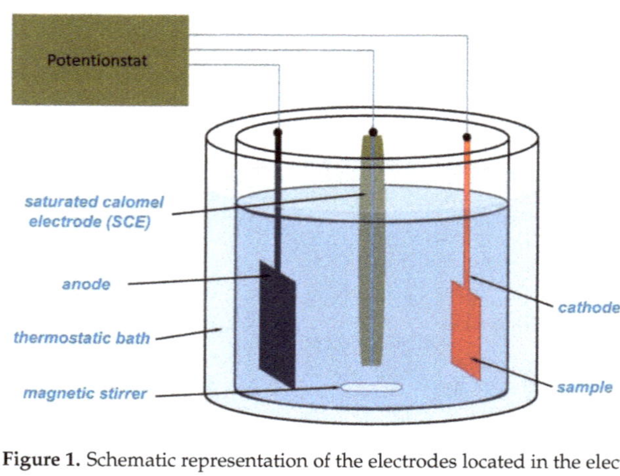

Figure 1. Schematic representation of the electrodes located in the electrochemical system.

A Philips (Amsterdam, The Netherlands) 1710D8 Advance diffractometer with CuKα radiation (instrumental broadening is 0.04° and λ = 1.54178 Å) was employed to determine X-ray diffraction (XRD, Philips, Amsterdam, The Netherlands) patterns in ZrO_2 layers. A Philips 515 scanning electron microscope (SEM, Philips, Amsterdam, The Netherlands) was used to observe the surface morphology of the deposited layers. The Zeta 20 3D optical profilometer (KLA, Milpitas, CA, USA) was used to determine the surface topography and average roughness (using the Zeta 3D_engr_1_8_5 software, KLA, Milpitas, CA, USA); the vertical resolution of the instrument was 1 nm. The measurements were performed at several different points on the surface of the ZrO_2 layers, each with an area of 6887 μm^2. The conductivity of the electrolyte was measured by BANTE 510 (Bante Instruments

Inc., Sugar Land, TX, USA). The spectra of transmittance and reflectance ((specular and diffuse component) and haze ratio) were obtained by a UV-VIS-NIR Shimadzu UV 3600 spectrophotometer (Shimadzu Scientific Instruments, Columbia, MD, USA).

3. Results and Discussion

The structure of the electrochemical layers of ZrO_2 deposited at different temperatures was determined by XRD analysis. The diffraction maxima are shown in Figure 2. The diffraction pattern of the substrate (SnO_2-covered glass) is shown with the black line for comparison. In the graph, we can distinguish six diffraction maxima related to zirconium dioxide; five of them are due to the monoclinic phase and one (which is the most intense) is due to the tetragonal phase. The diffraction maxima due to the SnO_2-covered glass substrate are marked with an asterisk (JCPDS No. 14–1445). The five maxima due to the monoclinic phase ((111), (112), (202), (013), and (−111)) are characteristic for ZrO_2 and are thoroughly described in the literature (JCPDS No. 37–1484) [28]. The diffraction maximum corresponding to the tetragonal direction (101) possesses the highest intensity (JCPDS No. 81–1544) [29]. This is relatively unusual for ZrO_2, because its crystallographic structure is monoclinic at deposition temperatures below 1170 °C. The explanation for this intense maximum may be related to epitaxial growth from the magnetron-sputtered SnO_2/glass substrate possessing tetragonal structure [30].

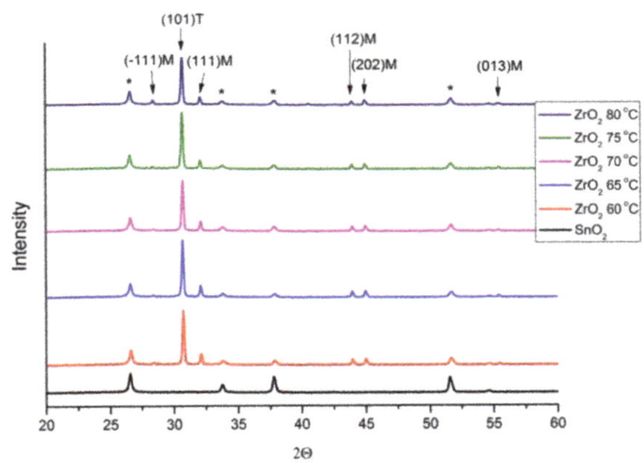

Figure 2. XRD patterns of the electrochemical ZrO_2 layers deposited on SnO_2 substrate at different temperatures. The diffraction maxima due to the SnO_2-covered glass substrate are marked with an asterisk (*).

Table 1 shows the positions of the diffraction maxima (2θ) and the values of the full width at half maximum (FWHM). From these diffraction maxima, the average grain size (D) was calculated using Scherrer's Equation (1) [31] for each crystallographic direction associated with a maximum:

$$D = (57.3 \times \lambda K)/(\beta \cos \theta), \quad (1)$$

where λ – (1.54178 Å) is the wavelength of CuK_1 radiation, θ is the Bragg angle, β is calculated from selected diffraction peak FWHM, and K is the particle shape factor (in most cases for particles with a predominantly regular shape it is assumed to be 0.9). There was no recognizable relationship between the calculation of the average grain size and the deposition temperature of the zirconia layers. We can note a slight decrease in the average grain size with the increase in the deposition temperature in the crystallographic directions with the greatest intensity of the diffraction maxima (101)T and (111)M. The missing results denoted with dash sign "-" in Table 1 were calculated with a large error due to the low intensity in the maxima in these directions, and for this reason, they were not shown.

Table 1. The position of the peak in the XRD patterns, 2θ, full width at half maximum (FWHM), β of 2θ, and the average grain sizes, D.

ZrO$_2$ Deposited on SnO$_2$ Substrate									
		(101)T			(111)M			(112)M	
	2Θ	FWHM	D, (nm)	2Θ	FWHM	D, (nm)	2Θ	FWHM	D, (nm)
60 °C	30.71	0.14	59	32.1	0.156	53	43.95	0.127	67
65 °C	30.68	0.151	54	32.08	0.162	51	43.95	0.230	37
70 °C	30.69	0.157	53	32.1	0.175	47	43.95	0.163	53
75 °C	30.68	0.160	52	32.07	0.165	50	43.93	0.146	59
80 °C	30.68	0.160	52	32.08	0.164	50	43.92	0.162	53
		(202) M			(013) M			(−111)M	
	2Θ	FWHM	D, (nm)	2Θ	FWHM	D, (nm)	2Θ	FWHM	D, (nm)
60 °C	44.99	0.190	45	55.41	0.116	77	-	-	-
65 °C	44.97	0.187	46	-	-	-	28.35	0.120	68
70 °C	44.99	0.254	34	-	-	-	28.67	0.162	51
75 °C	44.97	0.217	40	-	-	-	-	-	-
80 °C	44.96	0.197	44	-	-	-	28.40	0.118	70

Figure 3 shows the 3D optical profilometry in the electrochemical layers of ZrO$_2$ deposited at different temperatures. The RMS (root mean square) roughness values determined from the measurements are shown in Table 2. The RMS roughness in the SnO$_2$ substrate (magnetron sputtered on glass) is also presented for comparison. The calculations were made on the basis of data taken from three different sections (Sq^1, Sq^2, and Sq^3) of the surface of the samples, with each of them having an area of (97 μm × 71 μm = 0.69 mm^2). The results of optical profilometry did not show a noticeable relationship between the RMS roughness and the deposition temperature in the electrochemically deposited layers of ZrO$_2$. Here, the differences can be related to the different inhomogeneities in the layers, with the layer deposited at 65 °C (Figure 3c) possessing the greatest inhomogeneity. This inhomogeneity is not related to the substrate, as its RMS roughness is relatively constant and is at least twice lower than that in the deposited layers. The probable reason for the differences in the inhomogeneities at different temperatures is the different mobility of the ions in the solution, and the way they are attached to the substrate at each individual deposition temperature. Evidence of this is the electrolyte solution's changing conductivity with temperature; the data from the measurements in this parameter are shown in Table 3. Measurements show that as the electrolyte's temperature increases, the conductivity (σ) increases and resistivity (ρ) decreases.

Table 2. Average roughness in the ZrO$_2$ layers deposited at different temperatures on SnO$_2$ substrate.

	SnO$_2$ Substrate	Electrochemical ZrO$_2$				
		60 °C	65 °C	70 °C	75 °C	80 °C
Sq^1 (nm)	39	84	111	79	97	95
Sq^2 (nm)	34.5	82	119	76	95	87
Sq^3 (nm)	42.5	82	94	84	87	86
AVG (nm)	39 ± 2.7	83 ± 0.6	108 ± 8.2	79 ± 2.6	93 ± 3.4	89 ± 3

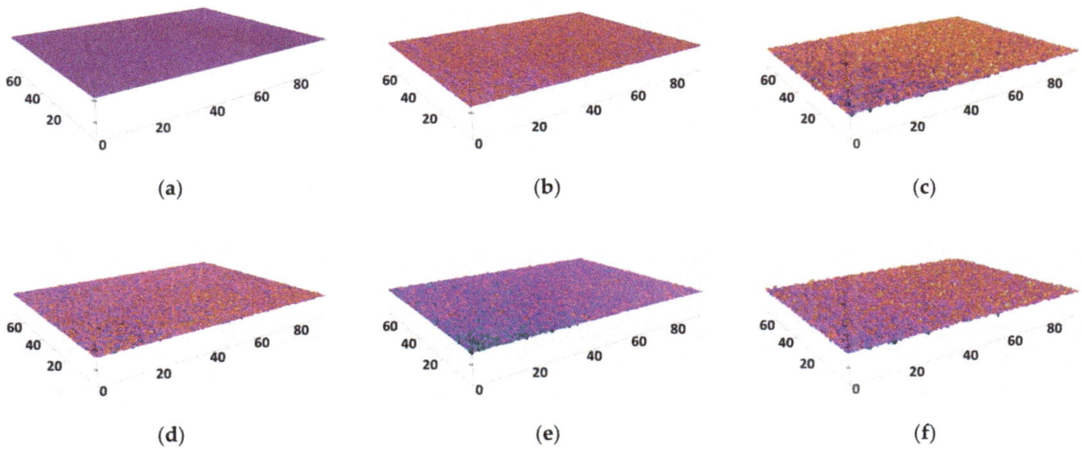

Figure 3. 3D optical profilometry images of SnO_2 substrate and electrochemically deposited ZrO_2 layers obtained at different temperatures for 20 min: SnO_2 substrate—(**a**); 60 °C—(**b**); 65 °C—(**c**); 70 °C—(**d**); 75 °C—(**e**); 80 °C—(**f**).

Table 3. Conductivity (σ) and resistivity (ρ) in electrochemical solutions measured at different temperatures.

	Electrochemical Solution				
Temperature	60 °C	65 °C	70 °C	75 °C	80 °C
σ–mS/cm	3.89	3.97	4.07	4.25	4.44
ρ–Ω/cm	257	252	246	235	225

The average thickness in the ZrO_2 layers deposited at different temperatures was measured with a 3D optical profilometer, and the results were: 60 °C—445 nm, 65 °C—517 nm, 70 °C—580 nm, 75 °C—595 nm, and 80 °C—480 nm.

Figure 4 shows the SEM micrographs of the magnetron-sputtered SnO_2 substrate on glass and the electrochemical layers of ZrO_2 deposited at different temperatures. The micrographs show that the zirconia layers have a predominantly granular structure with grain sizes of about 0.3–0.5 μm for the layers deposited at 60 °C (Figure 4b) and 1–2 μm for the layers deposited at 80 °C (Figure 4f). Here, in contrast to RMS's roughness, we can distinguish the dependence between the size of the grains located on the surface of the layers and the deposition temperature. With an increase in the deposition temperature, there was an evident increase in the size of the grains located on the surface of the electrochemical layers. The apparent shape of the grains of which the layers are composed is relatively regular. In Figure 4d, an onset of grain coalescence leading to the formation of larger crystallites is observed for the layer deposited at 70 °C. This layer appears to have the best morphology for the range of the deposition temperatures studied, and this can also be confirmed by the lowest-measured RMS roughness at this deposition temperature. This improved structure may be due to more favorable conditions, such as ion mobility and adhesion to the substrate, contributing to the formation of this type of structure.

Figure 4. SEM micrograph of SnO_2 substrate and ZrO_2 layers electrochemically deposited at different temperatures for 20 min: SnO_2 substrate—(**a**); 60 °C—(**b**); 65 °C—(**c**); 70 °C—(**d**); 75 °C—(**e**); 80 °C—(**f**).

The transmission, diffuse transmission, and haze ratio spectra of zirconia layers deposited at different temperatures are shown in Figure 5. For comparison, the spectra of the SnO_2 substrate are also shown in Figures 5 and 6. In the transmission spectra we observe a decrease in the values (related to the specular component of it) (Figure 5a) with increasing deposition temperature up to 75 °C. Further increases in temperature up to 80 °C show a slight increase in the values. In the diffuse transmission spectra (Figure 5b), an increase in the values is observed with increasing deposition temperature, in which case the layer deposited at the highest temperature (80 °C) has the highest values. The most likely explanation for the higher values in specular transmission for the layer deposited at 80 °C, compared to those deposited at 70 °C and 75 °C, is due to its lesser thickness, which comes from the different deposition rates depending on the electrolyte temperature. At higher deposition temperatures (from 80 °C), we obtained the formation of larger grains than those of the layers deposited at 70 and 75 °C (Figure 4) but located with a lower density on the surface; the lower density leads to higher values in the specular transmission of this layer. In the case of diffuse transmission, the grain size is also important. We observed the highest values in the spectra for the layer deposited at 80 °C because this layer has the largest grains, which leads to greater scattering. The values calculated for the haze ratio (Figure 5c) show the highest values in the visible region of the layers deposited at 75 °C, followed by those at 70 °C. This is because these layers have the lowest values of specular transmission, and the haze ratio represents the percentage ratio of the diffuse component related to the full transmission.

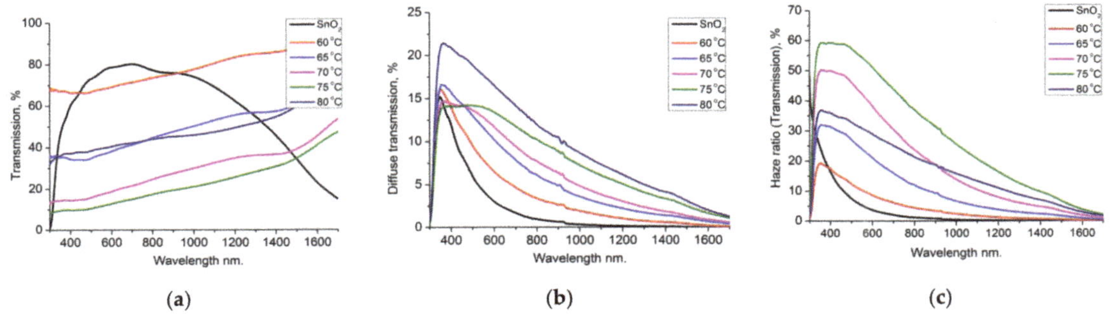

Figure 5. Transmittance (**a**), diffuse transmittance (**b**), and haze ratio (**c**) in ZrO_2 layers electrochemically deposited at different temperatures.

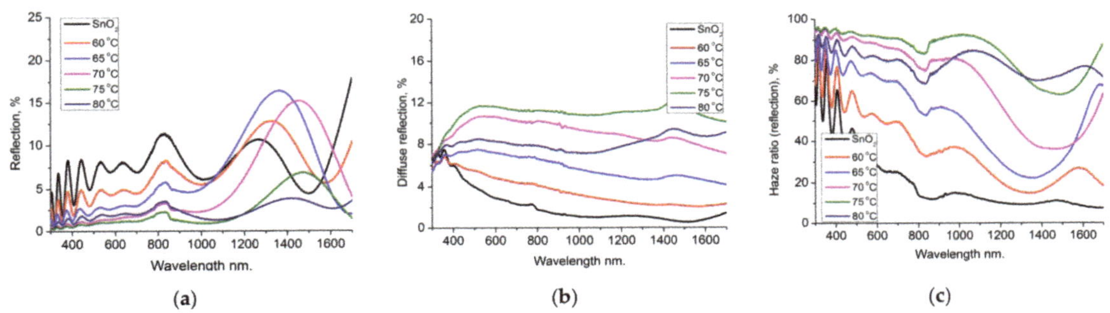

Figure 6. Reflectance (**a**), diffuse reflectance (**b**), and haze ratio (**c**) in ZrO_2 layers electrochemically deposited at different temperatures.

Figure 6 shows the optical spectra of reflection, diffuse reflection, and haze ratio of electrochemical ZrO_2 layers deposited at different temperatures. The spectra corresponding to the specular component of the reflection are shown in Figure 6a. Here, in general, the layers have very low values (less than 5%) in the visible range, and these values decrease with increasing deposition temperature to reach less than 1% for the layers deposited at 75 and 80 °C. Here, we observe strongly pronounced interference maxima that are a result of the SnO_2 substrate (magnetron sputtered on glass). This effect is observed when the layers are not thick and homogeneous enough and the substrate (which must have a flat parallel surface) is transparent to some extent and reflected through the layer. In the spectra of diffuse reflection (Figure 5b), the values are slightly higher than those of mirror reflection and reach about 10%–11% for the layers deposited at 70 and 75 °C. Unlike diffuse transmission, the layer deposited at 80 °C does not have the highest values, because in the reflection, apart from the size and shape of the grains, their surface density is also of great importance. For this reason, layers with more densely located grains on the substrate surface (such as layers deposited at 70 and 75 °C) have higher values in diffuse reflection. In diffuse reflection spectra, we do not observe interference maxima, because the SnO_2 substrate has negligibly small scattering values below 1%, which does not affect the electrochemical layers of ZrO_2. The values calculated for the haze ratio (Figure 6c) show high values in the visible range; as for the layers with a denser surface grain structure (those deposited at 70 and 75 °C), they reached 90%–95%. The "haze ratio" relationship is the percentage ratio of the components of diffuse transmission or reflection to the total transmission or reflection. The dependence of the optical spectra behavior on surface morphology is reported for ZrO_2 films obtained with different deposition methods, reflecting the peculiarities of the selected deposition method, the experimental conditions, and the specific structural and dielectric properties of the films arising from them. In [32], ZrO_2 thin films were obtained by magnetron sputtering onto optical grade quartz substrates. The crystallite size of the layers was in the range from 5 to 25 nm and was shown to significantly affect the optical characteristics of the thin films. The increasing crystallite size is associated with a random distribution of the grains, which makes the film surface rough, and as a result, an increase in light-scattering losses was observed. In [20], the average crystallite size of zirconium oxide films prepared by the magnetron-sputtering process at different argon partial pressures increases from 19 nm to 25 nm with increasing pressure. The minimum transmission values above 63% were observed for all films deposited at different pressures, and the values increased slightly with increasing argon partial pressure. This is explained by the thickness decreasing from 433 nm to 385 nm. In [33], homogenous and transparent nanocrystalline zirconia layers deposited on quartz substrates were produced by thermal oxidation using DC magnetron-sputtering techniques. It is shown that by increasing the annealing time from 60 to 240 min, the grains combine and form larger grains, in this manner changing the surface morphology from a pyramidal to cluster-type surface. By increasing the annealing time, the transparency of the zirconia films also increased. ZrO_2 films were also deposited by filtered cathodic vacuum arc [34]. The film structure is amorphous at room temperature and develops to polycrystalline upon heating the substrates to temperatures of 150 °C and above. Increasing the temperature leads to an increase in the surface's roughness. It was shown that changes in the film microstructure result in variations in their optical properties. For ZrO_2 thin films deposited through plasma-enhanced atomic layer deposition [17], it was shown that the transmittance increased monotonically with increasing wavelength. The lowest transmittance of 76.6% was measured at 300 nm, and the highest transmittance of 95.5% was measured at 800 nm (~30 nm; substrate: borosilicate glass; deposition temperature: 150 °C). ZrO_2 films [5] spin-coated with a simple water-based solution and cured with UV-A radiation = 330 nm for different times (40, 80, and 120 min) were compared with thermally annealed film (at 350 °C). The UV-A radiation-exposed films, in comparison to thermal treatment, exhibited a decrease in optical transmittances. The optical spectroscopy results demonstrated that increased doses of UV-A radiation improved the quality of films, in comparison with heat-treated films. In comparison with most of the above cited papers [5,17,32–34], we observed a greater size of grains and greater value in film roughness, leading to greater values in diffuse and haze ratio spectra. In any case, all these features depend on the applied deposition method, as well as the variety and specificity of working conditions.

4. Conclusions

The innovation approach we used to produce zirconia thin films by an environmentally compatible and quite inexpensive electrochemical deposition enriches and amplifies the existing data for essential properties of ZrO_2 layers grown by other deposition methods. We described our results on the impact of deposition temperature on the structural, morphological, and optical properties of nanostructured ZrO_2 films. The XRD patterns showed six (one in the tetragonal phase and five in the monoclinic phase) different diffraction maxima related to zirconium oxide, with no relationship between the deposition temperature and the average grain size. The calculations obtained from the optical profilometry did not show a relationship between RMS roughness and the deposition temperature in the electrochemical layers of ZrO_2, as the differences can be related to the different inhomogeneities in the layers. SEM micrographs revealed that the layers were composed mainly of granular structures, and the constituent formations had relatively regular spherical shapes. The size of these formations increased with increasing deposition temperature, and at 80 °C, they were about 1–2 μm. We also observed dependence in the optical spectra, as the values of the specular components of transmission and reflection decreased with increasing deposition temperature, and the values of the diffuse components and the haze ratio increased. It was demonstrated that the electrochemical method, in a relatively narrow and easily achievable temperature interval, allows the growth of diverse morphology and size nanostructures. Such layers of zirconia (with a wide energy band gap) with low values in specular and high values in diffuse transition and reflection in the visible and NIR range can be applied in some optoelectronic devices, such as background scatter layers in thin-film solar cells.

Author Contributions: Conceptualization, K.L.; methodology, K.L. and G.A.; software, K.L.; validation, G.A., R.G. and K.L.; formal analysis, G.A., R.G. and K.L.; investigation, K.L. and G.A.; resources, K.L.; data curation, K.L. and G.A.; writing—original draft preparation, K.L. and G.A.; writing—review and editing, G.A., R.G. and K.L.; visualization, K.L., R.G. and G.A.; supervision, K.L.; project administration, K.L.; funding acquisition, K.L. All authors have read and agreed to the published version of the manuscript.

Funding: This research paper is funded by the Bulgarian National Science Fund (BNSF) under the project КП–06–Н38/7 (12.2019).

Institutional Review Board Statement: Not applicable.

Informed Consent Statement: Not applicable.

Data Availability Statement: The authors confirm that the data supporting the findings of this study are available within the article.

Conflicts of Interest: The authors declare no conflict of interest. The funders had no role in the design of the study; in the collection, analyses, or interpretation of data; in the writing of the manuscript; or in the decision to publish the results.

References

1. Ravi Kumar, K.; Pridhar, T.; Sree Balajia, V.S. Mechanical properties and characterization of zirconium oxide (ZrO_2) and coconut shell ash (CSA) reinforced aluminium (Al 6082) matrix hybrid composite. *J. Alloys Compd.* **2018**, *765*, 171–179. [CrossRef]
2. Wang, L.; Cai, K.F.; Wang, Y.Y.; Yin, J.L.; Li, H.; Zhou, C.W. Preparation and characterization of tetragonal-ZrO_2 nanopowders by a molten hydroxides method. *Ceram. Int.* **2009**, *35*, 2499–2501. [CrossRef]
3. Wang, Y.; Zhou, X.; Liang, Z.; Jin, H. Characterization of Ultrasonic-Assisted Electrochemical Deposition of Ni-Co-ZrO_2. *Coatings* **2018**, *8*, 211. [CrossRef]
4. Lin, P.-C.; Lin, K.; Lin, Y.-H.; Yang, K.-C.; Semenov, V.I.; Lin, H.-C.; Chen, M.-J. Improvement of Corrosion Resistance and Biocompatibility of Biodegradable Mg–Ca Alloy by ALD $HfZrO_2$ Film. *Coatings* **2022**, *12*, 212. [CrossRef]
5. Bashir, A.; Farooq, M.; Malik, A.; Naseem, S.; Bhatti, A.S. UV-A Treatment of ZrO_2 Thin Films Fabricated by Environmental FriendlierWater-Based Solution Processing: Structural and Optical Studies. *Coatings* **2021**, *11*, 821. [CrossRef]
6. Melninkaitis, A.; Tolenis, T.; Mažulė, L.; Mirauskas, J.; Sirutkaitis, V.; Mangote, B.; Fu, X.; Zerrad, M.; Gallais, L.; Commandré, M.; et al. Characterization of zirconia– and niobia–silica mixture coatings produced by ion-beam sputtering. *Appl. Opt.* **2011**, *50*, 188–196. [CrossRef]
7. Pazhani, R.; Padma Kumar, H.; Varghese, A.; Moses Ezhil Raj, A.; Solomon, S.; Thomas, J.K. Synthesis, vacuum sintering and dielectric characterization of zirconia (t-ZrO_2) nanopowder. *J. Alloys Compd.* **2011**, *509*, 6819–6823. [CrossRef]
8. Deshmukh, S.B.; Bari, R.H.; Patil, G.E.; Kajale, D.D.; Jain, G.H.; Patil, L.A. Preparation and Characterization of Zirconia Based Thick Film Resistor as a Ammonia Gas Sensor. *Int. J. Smart Sens. Intell. Syst.* **2012**, *5*, 540–558. [CrossRef]
9. Deshmukh, S.B.; Bari, R.H. Nanostructured ZrO_2 Thin Films Deposited by Spray Pyrolysis Techniques for Ammonia Gas Sensing Application. *ILCPA* **2015**, *56*, 120–130. [CrossRef]
10. Hemalatha, E.; Gopalakrishnan, N. Synthesis of ZrO_2 nanostructure for gas sensing application. *Bull. Mater. Sci.* **2020**, *43*, 12. [CrossRef]
11. Bocanegra-Bernal, M.H.; Díaz de la Torre, S. Phase transitions in zirconium dioxide and related materials for high performance engineering ceramics. *J. Mater. Sci.* **2002**, *37*, 4947–4971. [CrossRef]
12. Husain, A.; Ahmad, S.; Mohammad, F. Thermally stable and highly sensitive ethene gas sensor based on polythiophene/zirconium oxide nanocomposites. *Mater. Today Commun.* **2019**, *20*, 100574. [CrossRef]
13. Hemalatha, E.; Gopalakrishnan, N. Gas sensing performances of pure and Cu-doped ZrO_2 nano structures. *Appl. Phys. A* **2019**, *125*, 493. [CrossRef]
14. Pai, A.R.; Nair, B. Synthesis and characterization of a binary oxide ZrO_2–TiO_2 and its application in chlorophyll dye-sensitized solar cell with reduced graphene oxide as counter electrodes. *Bull. Mater. Sci.* **2015**, *38*, 1129–1133. [CrossRef]
15. Ziębowicz, A.; Sambok-Kiełbowicz, A.; Walke, W.; Mzyk, A.; Kosiel, K.; Kubacki, J.; Bączkowski, B.; Pawlyta, M.; Ziębowicz, B. Evaluation of Bacterial Adhesion to the ZrO_2 Atomic Layer Deposited on the Surface of Cobalt-Chromium Dental Alloy Produced by DMLS Method. *Materials* **2021**, *14*, 1079. [CrossRef] [PubMed]
16. Minh, N.Q. Ceramic fuel cells. *J. Am. Ceram. Soc.* **1993**, *76*, 563–588. [CrossRef]

17. Oh, J.; Seo, G.; Kim, J.; Bae, S.; Park, J.-W.; Hwang, J.-H. Plasma-Enhanced Atomic Layer Deposition of Zirconium Oxide Thin Films and Its Application to Solid Oxide Fuel Cells. *Coatings* **2021**, *11*, 362. [CrossRef]
18. Lim, H.S.; Ahmad, A.; Hamzah, H. Synthesis of zirconium oxide nanoparticle by sol-gel technique. *AIP* **2013**, *1571*, 812. [CrossRef]
19. Kuo, D.H.; Chien, C.H. Growth and properties of sputtered zirconia and zirconia–silica thin films. *Thin Solid Film.* **2003**, *429*, 40–45. [CrossRef]
20. Patel, U.S.; Patel, K.H.; Chauhan, K.V.; Chawla, A.K.; Rawal, S.K. Investigation of Various Properties for Zirconium Oxide Films Synthesized by Sputtering. *Procedia Technol.* **2016**, *23*, 336–343. [CrossRef]
21. Sawka, A. Chemical Vapour Deposition of Scandia-Stabilised Zirconia Layers on Tubular Substrates at Low Temperatures. *Materials* **2022**, *15*, 2120. [CrossRef] [PubMed]
22. Tcheliebou, F.; Boulouz, M.; Boyer, A. Preparation of fine-grained MgO and Gd_2O_3 stabilized ZrO_2 thin films by electron beam physical vapor deposition co-evaporation. *J. Mater. Res.* **1997**, *12*, 3260–3265. [CrossRef]
23. Prieto-López, L.O.; Yubero, F.; Machorro, R.; De La Cruz, W. Optical properties of Zr and ZrO_2 films deposited by laser ablation. *Microelectron. J.* **2008**, *39*, 1371–1373. [CrossRef]
24. Wang, S.; Shen, J. Fabrication of sol–gel derived ZrO_2 thin film for HR coatings via rapid thermal annealing process. *J. Sol-Gel Sci. Technol.* **2013**, *67*, 339–343. [CrossRef]
25. Tang, L.; Maruyama, H.; Han, T.; Nino, J.C.; Chen, Y.; Zhang, D. Resistive switching in atomic layer deposited HfO_2/ZrO_2 nanolayer stacks. *Appl. Surf. Sci.* **2020**, *515*, 146015. [CrossRef]
26. Gal-Or, L.; Silberman, I.; Chaim, R. Electrolytic ZrO_2 Coatings: I. Electrochemical Aspects. *J. Electrochem. Soc.* **1990**, *138*, 1939. [CrossRef]
27. Lopes, N.I.A.; Freire, N.H.J.; Resende, P.D.; Santos, L.A.; Buono, V.T.L. Electrochemical deposition and characterization of ZrO_2 ceramic nanocoatings on superelastic NiTi alloy. *Appl. Surf. Sci.* **2018**, *450*, 21–30. [CrossRef]
28. Jafarpour, M.; Rezapour, E.; Ghahramaninezhad, M.; Rezaeifard, A. A novel protocol for selective synthesis of monoclinic zirconia nanoparticles as a heterogeneous catalyst for condensation of 1,2-diamines with 1,2-dicarbonyl compounds. *New J. Chem.* **2014**, *38*, 676–682. [CrossRef]
29. Sharifi, H.; Divandari, M.; Khavandi, A.; Idris, M.H. Effect of Al powder and silica sol on the structure and mechanical properties of Al_2O_3-ZrO_2 foams. *Acta Metall. Sin.* **2010**, *23*, 241–247. [CrossRef]
30. Lovchinov, K.; Slavov, L.; Alexieva, G.; Ivanov, P.; Marinov, G.; Gergova, R.; Strijkova, V.; Babeva, T. Study of ZrO_2 nanolayers deposited electrochemically on different conductive substrates. *Mater. Sci. Semicond. Process.* **2021**, *131*, 105843. [CrossRef]
31. Mass, J.; Bhattacharya, P.; Katiyar, R. Effect of high substrate temperature on Al-doped ZnO thin films grown by pulsed laser deposition. *Mater. Sci. Eng. B.* **2003**, *103*, 9–15. [CrossRef]
32. Ramana, C.V.; Vemuri, R.S.; Fernandez, I.; Campbell, A.L. Size-effects on the optical properties of zirconium oxide thin films. *Appl. Phys. Lett.* **2009**, *95*, 231905. [CrossRef]
33. Hojabri, A.; Pourmohammad, S. Optical Properties of Nano-Crystalline Zirconia Thin Films Prepared at Different Post-Oxidation Annealing Times. *Acta Phys. Pol. A* **2016**, *129*, 647–649. [CrossRef]
34. Zhao, Z.W.; Tay, B.K.; Huang, L.; Yu, G.Q. Study of the structure and optical properties of nanocrystalline zirconium oxide thin films deposited at low temperatures. *J. Phys. D Appl. Phys.* **2004**, *37*, 1701–1705. [CrossRef]

coatings

MDPI

Review

A Review on Sustainable Manufacturing of Ceramic-Based Thin Films by Chemical Vapor Deposition (CVD): Reactions Kinetics and the Deposition Mechanisms

M. Sabzi [1], S. H. Mousavi Anijdan [2,3,*], M. Shamsodin [4], M. Farzam [5], A. Hojjati-Najafabadi [6], P. Feng [6], N. Park [7] and U. Lee [8,9]

1 School of Metallurgy and Materials Engineering, Iran University of Science and Technology (IUST), Tehran, Iran
2 Department of Materials Engineering, Science and Research Branch, Islamic Azad University, Tehran, Iran
3 Department of Advanced Materials & Processing, Research and Development of Engineering Materials Research Center, Science and Research Branch, Islamic Azad University, Tehran, Iran
4 Department of Mining and Metallurgical Engineering, Amirkabir University of Technology (Tehran Polytechnic), Tehran, Iran
5 Department of Technical Inspection Engineering, Petroleum University of Technology, Abadan, Iran
6 School of Materials Science and Physics, China University of Mining and Technology, Xuzhou 221116, China
7 School of Materials Science and Engineering, Yeungnam University, 280 Daehak-ro, Gyeongsan 38541, Republic of Korea
8 POSCO Technical Research Laboratories, Gwangyang 57807, Republic of Korea
9 BISTEP Evaluation & Analysis of Regional Innovation Program Division, Busan 48058, Republic of Korea
* Correspondence: hashemmousavi@gmail.com or hashemmousavi@srbiau.ac.ir

Citation: Sabzi, M.; Mousavi Anijdan, S.H.; Shamsodin, M.; Farzam, M.; Hojjati-Najafabadi, A.; Feng, P.; Park, N.; Lee, U. A Review on Sustainable Manufacturing of Ceramic-Based Thin Films by Chemical Vapor Deposition (CVD): Reactions Kinetics and the Deposition Mechanisms. *Coatings* **2023**, *13*, 188. https://doi.org/10.3390/coatings13010188

Academic Editor: Ben Beake

Received: 18 November 2022
Revised: 23 December 2022
Accepted: 9 January 2023
Published: 14 January 2023

Abstract: Chemical vapor deposition (CVD) is a process that a solid is formed on a substrate by the chemical reaction in the vapor phase. Employing this technology, a wide range of materials, including ceramic nanocomposite coatings, dielectrics, and single crystalline silicon materials, can be coated on a variety of substrates. Among the factors influencing the design of a CVD system are the dimensions or geometry of the substrate, substrate temperature, chemical composition of the substrate, type of the deposition process, the temperature within the chamber, purity of the target material, and the economics of the production. Three major phenomena of surface reaction (kinetic), diffusion or mass transfer reaction, and desorption reaction are involved during the CVD process. Thermodynamically, CVD technology requires high temperatures and low pressures in most systems. Under such conditions, the Gibbs free energy of the chemical system quickly reaches its lowest value, resulting in the production of solids. The kinetic control of the CVD technology should always be used at low temperatures, and the diffusion control should be done at high temperatures. The coating in the CVD technology is deposited in the temperature range of 900–1400 °C. Overall, it is shown here that by controlling the temperature of the chamber and the purity of the precursors, together with the control of the flow rate of the precursors into the chamber, it is possible to partially control the deposition rate and the microstructure of the ceramic coatings during the CVD process.

Keywords: chemical vapor deposition (CVD); ceramic composite coatings; layer conditions; deposition mechanism; reaction kinetics

1. Introduction

The chemical vapor deposition (CVD) technique was originally developed, as a new manufacturing process, and to effectively fabricate a range of engineering products such as nanocomposite ceramic coatings, critical components in numerous industrial sectors including semiconductor, ceramic, mining, etc. [1,2]. Nowadays, the CVD technique has gone far beyond its original scope, particularly in the semiconductor and microelectronics industries, due to extensive research in various fields. The reasons for the uniqueness of the CVD technique include the ability to produce highly versatile layers, the possibility to apply nanocomposite ceramic coatings on metals, the ease of semiconductors fabrication, and the opportunity to fabricate layers with organic and inorganic compounds. The layers created by the CVD technique are usually in the form of crystalline or amorphous with versatile properties, which can be acquired by controlling the production parameters [2,3]. In fact, the CVD technique is part of ceramic coating deposition techniques that have been advancing relatively fast in recent years. Generally, in the CVD process, a solid layer is produced on the substrate by a chemical reaction in the vapor phase. Historically, soot formation due to the incomplete oxidation and burning of firewood is perhaps the oldest example with close proximity to the CVD technique [4,5].

Among the different layer deposition methods for ceramic nanocomposite coatings, the CVD technique is the most important one in industrial production due to its relatively low cost of fabrication and also due to its high production efficiency. The CVD technique is a process that involves the decomposition or chemical reaction of gaseous reactants under heat, light, and plasma. Such a decomposition is used to form a stable solid layer. The CVD can produce highly purified stand-alone materials or coatings whose characteristics can be controlled down to atomic or nanometer scale [6–8]. In addition to that, this process can produce monolayer materials, multilayer materials, composite materials, nanostructure pieces, and coatings with a specific grain size structure that contains excellent dimensional stability [8,9].

The applications of CVD technology have significantly expanded over the past few years. Relying on the deposition aspects of the process, CVD technology has found many applications in materials science and engineering. Varying application of the CVD technology from extraction to precipitation has made technology a very compelling and important coating technique. For example, the CVD technology is extensively used to produce thin film semiconductor and nanocomposite ceramic coatings with much improved surface properties such as abrasion protection, corrosion/oxidation resistance, chemical reactions, thermal stress, and neutron absorption. In historical terms, the early 1970s was the period when CVD technology achieved a great triumph in manufacturing electronic semiconductors, and protective coatings for electronic circuits. These successes have resulted in the rapid expansion of CVD technology to other areas such as manufacturing processes of ceramic coatings. Of particular arena of the expansion of the technology was in advanced ceramics for manufacturing of high temperatures tools such as high temperature ceramic coatings, turbine blade coatings, fiber-reinforced composites, and solar cells, to name a few [2,6]. Nowadays, CVD technology is highly useful and is getting greater importance, in manufacturing strategic components in the fields of aerospace, military, aviation, nuclear, and general materials engineering.

CVD technology is often used to produce coatings, powders, fibers, and unified components. It is possible to produce most metals, some non-metals such as carbon and silicon as well as many components including carbides, nitrides, oxides, and intermetallic phases, etc. Technology is one of the essential factors in producing semiconductors and other electronic components, tool coating, bearings, parts resistance to wear and corrosion, and optical instruments [9,10]. Using this process, a variety of nanostructures such as quantum dots, ceramic nanostructures, carbides, carbon nano-pipes, and even diamonds are producible, which is one of the main advantages of the process. Due to the high speed of the process, it is quite possible to prepare different industrial nanostructures using the process in a way that even single-wall nanopipes are industrially produced by this method. Diverse pre-materials can be used during the process which is because of high the temperature of the reaction. Another advantage of this process is the absence of many extra side materials. The extra materials are mostly gaseous and can be separated from the main materials [8–12].

Varieties of coatings or heat treatment procedures are used to compensate for the shortcomings in metals and alloys (i.e., low corrosion resistance, and poor wear performance) [10–18]. Often, the manufacturing process is modified through the tweaking of the production process to achieve the desired properties [19–25]. The specifications of a large number of articles related to ceramic coatings prepared by the CVD process are reported in Table 1.

Table 1. Example publications demonstrating the range of ceramic coatings prepared by the CVD technology.

No.	Reference Specifications
1	High-Temperature Oxidation Behavior of CVD-SiC Ceramic Coating in Wet Oxygen and Structural Evolution of Oxidation Product: Experiment and First-Principle Calculations [26].
2	Effect of PyC-SiC Double-Layer Interface on Ablation Behaviour of Impacted CVD-SiCnws/HfC Coating [27].
3	Influence of Crystallite Morphology on the Ablative Behaviors of CVD-TaC Coatings Prepared on C/C Composites Beyond 2100 °C [28].
4	Effect of Microstructure on the Ablation Behavior and Mechanical Properties of CVD-HfC Coating [29].
5	Complex Geometry Macroporous SiC Ceramics Obtained by 3D-printing, Polymer Impregnation and Pyrolysis (PIP) and Chemical Vapor Deposition (CVD) [30].
6	Low-Temperature SiO_2Film Coatings onto Cu Particles Using the Polygonal Barrel-plasma Chemical Vapor Deposition Method [31].
7	Fabrication of Robust Ceramic Based Super hydrophobic Coating on Aluminum Substrate via Plasma Electrolytic Oxidation and Chemical Vapor Deposition Methods [32].
8	Improvement of Ablation Resistance of CVD-HfC/SiC Coating on Hemisphere Shaped C/C Composites by Introducing Diffusion Interface [33].
9	Hard Turning Performance Evaluation Using CVD and PVD Coated Carbide Tools: A Comparative Study [34].
10	Growth Mechanism and Ablation Behavior of CVD-HfC Coating on the Surface of C/C Composites and CVD-SiC Coating [35].
11	Preparation and Ablation Resistance of ZrC Nanowires-Reinforced CVD-ZrC Coating on Sharp Leading Edge C/C Composites [36].
12	Fabrication of Porous SiC Nanostructured Coatings on C/C Composite by Laser Chemical Vapor Deposition for Improving the Thermal Shock Resistance [37].
13	CVD Synthesis of Nanometer SiC Coating on Diamond Particles [38].
14	Microstructure Evolution and Growth Mechanism of Si-$MoSi_2$Composite Coatings on TZM (Mo-0.5Ti-0.1Zr-0.02 C) Alloy [39]
15	Optimisation of Spray-Mist-Assisted Laser Machining of Micro-Structures on CVD Diamond Coating Surfaces [40].

Given the importance of CVD technology and its rapid expansion in many sectors, this study was conducted to review the CVD deposition mechanism of ceramic nanocomposite coatings on metal substrates.

2. Principles of the CVD Process

In most common devices, the CVD technique involves the flow of gases or precursor gases into a chamber. There are one or more hot substrates in the chamber, where the coating is going to be deposited. In this process, chemical reactions occur on or near the hot substrate. The result is a deposition in the form of a crystalline or amorphous thin film or a combination of both products on the substrate [41,42]. One of the disadvantages of this process is the production of chemicals. Waste materials and byproducts come-out of the chamber along with the unreacted precursor gases [40,43]. The deposition is typically carried out in the temperature range of 900–1400 °C. Unlike the physical vapor deposition (PVD) process, which involves processes such as evaporation, dispersion, and sublimation, the CVD process involves chemical reactions in the precursor (or between the precursors) [41–45].

The CVD technique is a process in which a solid is produced by a chemical reaction in the vapor phase. This reaction is performed on the substrate, which is either cold or often preheated. The solid that is produced in this process can be a coating, a powder, or a single crystal material. Materials with different properties can be produced by varying the parameters of the CVD process such as substrate type, substrate temperature, the chemical composition of the mixtures of the reactant gases, and the total pressure of the gas that flows, etc. [46,47]. The materials produced are very pure. It is mainly in the form of nanostructures and is widely used in the manufacturing of semiconductor pieces. A wide range of products including ceramic nanocomposite coatings, dielectrics, single-crystalline silicone, polymeric materials, and metals can be coated to the substrate by such a layering process [46–50]. One of the oldest applications of the CVD process is the manufacturing of different pigments in the industry. Usually, TiO_2, SiO_2, Al_2O_3, Si_3N_4, and even carbon black powders can be made in nano or micron sizes by the CVD process. Due to the nature of the CVD process, in some circumstances, this method is used to precipitate some pure metals on the substrate. Similar to the PVD process, several parameters influence the quality of layer(s) that is produced in the CVD process. Of the most important parameters are substrate temperature, reactant concentration, gas pressure, and gas flow rate [49–51].

The most important part of the CVD process is the chemical reactions occurring in the chamber and between the reactant gases as well. As a result of such reactions, the desired solid is deposited in a crystalline or amorphous form on the substrate. Depending on the amount of pressure, the chemical reactions during the CVD process may be in the CVD atmospheric pressure, known as APCVD, or at a low pressure CVD known as LPCVD, in the approximate range of 0.1 to 25 torr. Nonetheless, the process might be performed in the high pressure CVD (HPCVD). Moreover, one or more gases are pressurized at low pressures in the CVD method. They are then compressed and mixed at or near the substrate surface under controlled conditions, where the reaction occurs between the gases [52,53]. However, due to the nature of the gases, and the possibility of the production of highly toxic or degradable products during the deposition process, attention must be paid to the selection of the reactive gases [54–57]. In general, the following are most of the common reactions that occur, during the CVD process, for the deposition of nanostructured coatings or thin films [57–60]:

(a) Decomposition of reactive gases.
(b) Gas combination.
(c) Gas hydrolysis.
(d) Gas oxidation.
(e) Reduction of some gases.

Usually, the formation of solid particles in the gaseous phase should be avoided. This is due to the formation of nonuniform layer thickness in addition to the raw materials wasting. Unwanted particles are also formed in the solid film. However, the formation of solid particles in the gas phase can be used under certain laboratory conditions. For example, gas-phase germination, and particle growth control are the most important aspects of the growth processes for the manufacturing of nanocomposite coatings, nanoparticles, or nano-powders. The particle size range is controlled by the number of nuclei formed in the reactor as well as the rate of particle contraction. Figure 1 shows an example of a typical CVD system in which reactive gases, commonly referred to as precursor gases, enter the chamber at an appropriate temperature. As the gases pass through the reactor, they come in contact with the hot substrate. They then react with the substrate and a solid layer of ceramic coating characteristic (for example, Al_2O_3, Cr_2O_3, or SiC) is deposited on the substrate. It should be noted that a neutral gas, such as argon, is commonly used in the CVD process as a diluent [59,60]. As mentioned earlier, the deposition temperature and pressure are usually the two limiting factors in this process.

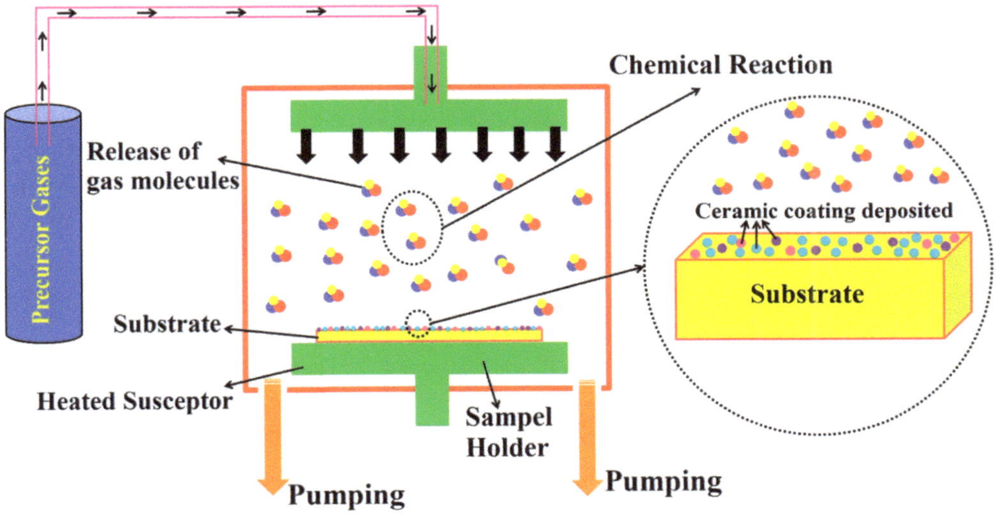

Figure 1. An example of a CVD system for the deposition of ceramic nanocomposite coatings.

3. Deposition Mechanism in the CVD Process

It is generally recognized that three major phenomena (surface reaction (kinetic), diffusion or mass transfer reaction, and desorption reaction) are involved during the CVD process [58–61]. Figure 2 shows a schematic of the main steps happening during the CVD process. The first step is to transfer the reactants (precursor gases) to the area that is supposed to be deposited. This is carried out by the convective current applied in the process. The second and fifth steps are dependent on each other through the stoichiometric relationships of the reactants due to the influence of the infiltration rate of the reactants. The third and fourth steps are very complex. They may involve surface reactions, simultaneous adsorption and desorption (chemical adsorption and physical desorption), and nucleation processes on the substrate. Commonly, the slowest step determines the overall speed of the chemical reactions. The rate of decomposition of the raw materials on the substrate increases if the substrate temperature in the chamber is much higher than the reactant decomposition temperature. Under such a circumstance, the growth rate of the deposited coatings/layers is controlled by the rate of transfer of the mass of the reactants on the substrate [57–64]. However, if the substrate temperature in the chamber is much higher than the reaction temperature of the gases, the rate of reactions that occur is limited by the temperature dependent heterogeneous nucleation rate, and the effect of desorption/kinetics. Therefore, the initial reaction rate in the chamber is controlled by the kinetics of the reaction [63–67]. In this case, if the amount of the raw materials is appropriate, the growth rate is independent of the temperature of the chamber and is only exponentially dependent on the temperature of the raw materials. Therefore, it can be stated that in the CVD process, the temperature is the rate controller at lower temperatures, and diffusion is the rate controller at higher temperatures [66–69].

Generally, the CVD techniques are performed at very high temperatures and at low pressures (to achieve high deposition rates). Therefore, the reaction rate is not somehow temperature-dependent in this process. However, the reaction rate in the chamber is highly dependent on the flow rate of the raw materials or the interface between the gases produced and the substrate. At very high temperatures, where the amount of supersaturation is high and the reactant gases are very hot, the nucleation rate is homogeneous in the CVD process, thereby precipitating solid particles on the hot substrate. With the progression of the deposition time and the consumption of some of the precursors, the raw material decreases, and subsequently the rates of the nucleation and growth on the substrate decrease as such. Therefore, it can be stated that the amount of raw material injected into the chamber controls the reaction rate. On the other hand, the rate of chemical reactions usually increases with increasing the pressure of the system. Overall, nucleation in the gaseous stage is homogeneous, whereas it is heterogeneous on the substrate surface [69–74].

Generally, the basic steps in the CVD process can be classified as follows [71–75]:

(a) Heat transfer and diffusion of the reactants from the gas reservoir to the reaction zone.
(b) Occurrence of chemical reactions in the gas phase to produce reactive samples and byproducts.
(c) Transfer of primary reactants and their products onto the metallic or nonmetallic substrate.
(d) Chemical adsorption and the diffusion of these products on the metallic or nonmetallic substrate.
(e) Inhomogeneous reactions by the surface resulting in the formation of ceramic nanocomposite coatings on the metallic or nonmetallic substrate.
(f) Heat transfer and outflow of the by-products outside the reaction chamber (deposition layer zone).

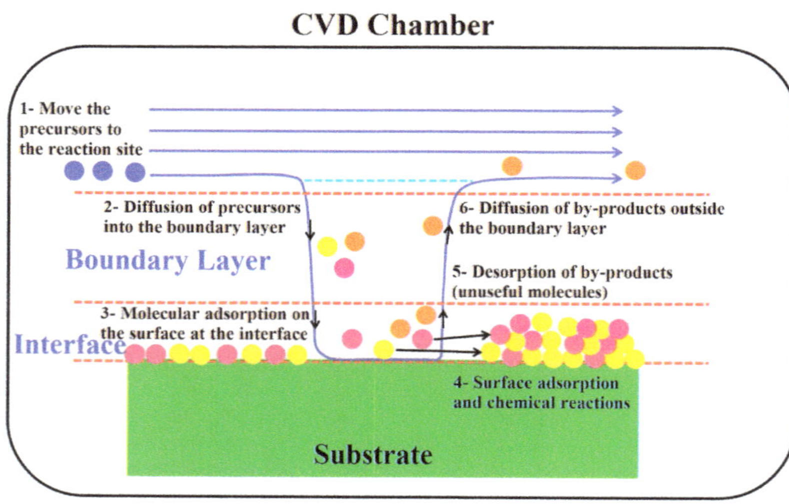

Figure 2. Schematic of the main steps of the CVD process.

The nucleation stage of the CVD process involves the formation of critical-sized clusters (or nuclei with the critical size) that subsequently grow to become stable crystals. Figure 3 shows the three main models that are considered, for the growth process. The dimension or geometry of the substrate, substrate temperature, chemical composition of the substrate, type of the deposition process, the temperature of the chamber, the purity of the target material, and the economic factors of production influence the design of the CVD system [75–77]. Therefore, it can be said with high certainty that any changes in the manufacturing condition to produce critical engineering components can alter the microstructural, mechanical, and corrosion properties of these components [78–86]. In fact, the liquid or solid precursor is heated to achieve a gaseous phase in the CVD process. The generated gas is then purged in the CVD reactor at a suitable speed. When the specific thickness of the film or the nanocomposite coating or the combination of several layers varies, the rate of gas evaporation should be controlled. Nowadays, a digital flow controller or pressure-flow controller is usually used in advanced CVD systems to achieve gas evaporation rate control. In a sense, liquid precursors are usually easier to use compared to solid precursors. This is due to the fact that the heat transfer and surface area available to the solid precursors are lower than the liquid precursors [85–89].

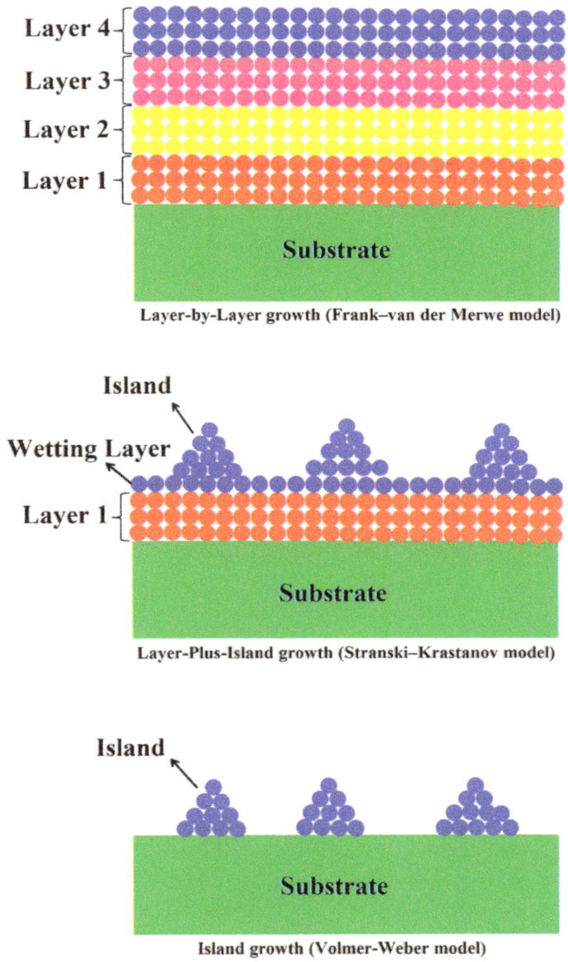

Figure 3. Basic models available for the growth process and layering on the substrate.

4. Kinetics of Reactions and Coating Production in the CVD Process

Thermodynamically, in most systems, the CVD process requires high temperatures and low pressures. Under such conditions, the Gibbs free energy of the chemical system quickly reaches its lowest value, resulting in the stability and production of solid products [90,91]. The first step in understanding this process is the application of the concepts of thermodynamics together with the concepts of gas transport, reaction kinetics, nucleation, and growth of the particles that were precipitated. This information can be very valuable in choosing a suitable system for the reaction. The CVD process usually consists of the following five steps [92–94]:

1—Entrance of the reactant gases into the reactor.

2—Diffusion of the gases through a boundary layer with the help of convective flow.

3—Connection of the gases with the substrate at the gas–substrate interface.

4—Precipitation operations on the substrate.

5—The diffusion of the products that were produced through the boundary layer on the substrate.

The concept of boundary layer resistance against infiltration was first proposed by Noise and Whitney. It was developed by Nernst in subsequent years. The thickness of the boundary layer does not change significantly with the deposition time. However, it was shown in experimental systems that in high-pressure systems, the thickness of this boundary layer is altered due to its suitable penetration speed. Generally, if the mass transfer rate is much greater than the kinetic rate constant of the products, and the pressure of the system at the boundary layer as well, the reaction is said to be controlled by the kinetics. However, if the opposite is the case, the reaction is controlled by diffusion. If there are no thermodynamic constraints, the kinetic control of the CVD process should always be used at low temperatures, and at diffusion control at high temperatures. On the other hand, the reaction rate constant is exponentially related to temperature [94–97]. Figure 4 shows the importance of the relationship between the kinetic factor (growth rate) and the diffusion during the layering process of the CVD process.

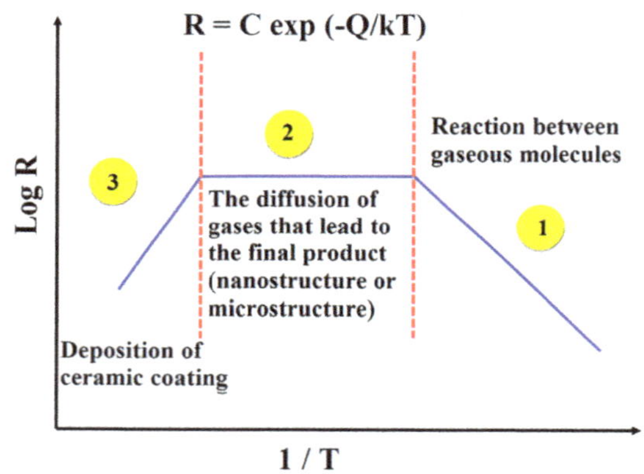

Figure 4. The relationship between kinetic factor (growth rate) and diffusion with temperature during the deposition (ideal indicator chart for coatings deposited in the CVD process).

Typically, at relatively low temperatures in the CVD process, the rate of reaction is limited by the strong dependence of the heterogeneous nucleation reaction rate on temperature as well as the desorption and kinetics effects. Thus, the initiation and rate of the initial reaction are controlled by kinetics. In this situation of the CVD process, the growth rate is not sensitive to the raw material flow rate except for the low flow of the raw material. However, the growth rate has an exponential relationship with temperature. At higher temperatures where the reaction rate is high, the diffusion of the raw materials and the gaseous products to the boundary layer is the determining factor of the reaction rate. As a result, the initial speed of the reaction is controlled by diffusion. In fact, in such a scenario, the reaction rate is not temperature-dependent. However, it is highly dependent on the flow rate of the raw materials or the boundary layer. At very high temperatures where the supersaturation is high and the reactant gases are very hot, the nucleation reaction is homogeneous. Therefore, the solid particles precipitate on the substrate. Consequently, due to the reduction of the raw materials (by precipitation of the solid particles on the substrate), the growth rate decreases on the substrate in a way that the growth temperature increases [97–100].

Moreover, similar changes occur in the mechanism of temperature-constant and pressure-variable systems. Under abnormal conditions, proper mass transfer is independent of the pressure of the system. However, the rate of chemical reaction usually increases with increasing the pressure of the system. Therefore, the reaction phase can be limited when the pressure of the system is low. While the diffusion stage can be controlled at high pressures. At temperatures that are suitable for the thermal degradation of the materials, the homogenous nucleation of solid particles in the gas phase is more convenient at low pressures in hot wall reactors compared to high pressures. This is due to the short free scanning length of the reaction radicals at high pressures. Generally, nucleation is homogeneous in the gas phase. While nucleation is heterogeneous on the substrate. It seems that the nucleation process involves the formation of critical-sized clusters where stable crystals form in the latter stage [98–102]. In this process, three main models are considered for the growth process, which is shown in Figure 3.

It is important to note that one can achieve layer-by-layer growth (Frank–van der Merwe growth model) under some special conditions. If the correlation between the lattice parameter of the structure and the film grid parameter is too high in a way that the mismatch is too low, three-dimensional clusters are formed in the Volmer–Weber growth model. As the mismatch between the lattice of the substrate and the film grid increases, the 3D clusters have a suitable energy level so long as the relative potential energy is greater than zero. As a result, intermediate films named Stranski–Krastanov are formed. Initially, a few single-layers are created and then clusters are then formed. When the energy between the film–substrate interface decreases, the film is separated from the substrate and the clusters are formed on the surface of the first precipitated single-layer [86–90]. As stated previously, the dimensions, shape, composition of the substrate, the type of the precipitation process, thermodynamics of the deposited materials, and economic factors are influential on the CVD process. The liquid or solid precursor enters the gas through a boiling evaporator. Then, the gas that is produced is sent into the reactor. The evaporation rate should be controlled when a particular thickness of a film or layer or multiple compositions with a specific molar ratio is desired. In advanced CVD systems, current controllers or pressure–current controllers are commonly used. And liquid precursors are easier to use compared with solid precursors since heat transfer and the available surface area of the liquid are higher than that of the solids [103–106].

In contrast to thermodynamic consideration that talks about the possibility of a reaction, kinetics talks about the chemical conversion over the unit of chemical reaction time [107,108]. The overall reaction is subdivided into several reactions, where the speed of the CVD process is determined by the rate of the slowest reaction. Therefore, the rate of this slowest reaction affects the deposition rate as well. When empirically determining the deposition rate, an ideal index diagram of the CVD process is obtained. Such an ideal diagram is shown in Figure 4. It can be observed that this diagram has three index zones. At low temperatures, the deposition process is determined by the reaction performed on the surface and follows the Arrhenius equation. This reaction-controlled region (Zone 1) is suitable for creating uniform coatings on complex geometries. At higher temperatures, the reaction speed is so high that the deposition process is carried out by particle transport (Zone 2). However, at even higher temperatures, powder formation usually occurs (Zone 3). For economic reasons, most of the CVD reactors in operation are restricted to Zone 2 [109–113]. When a uniform composite coating layer on a substrate with a complex geometry is the first major consideration, CVD processes generally work in region 1 of the diagram in Figure 4.

The mass transfer of gases to the substrate in zone 2 determines the layering velocity in the CVD process. A CVD process is accomplished by a complex and dynamic fluid flow pattern. For example, gas mixtures flow through pipes, valves, and chambers, while at the same time they are exposed to high temperatures. This temperature is different than the pre-mixing temperature. Then, the gas mixture (nanocomposite coating) reacts with the metal substrate. This reaction is a heterogeneous one due to the phase transformation from gaseous to solid state during the CVD process. Figure 5 shows the steps that take place for the deposition of ceramic nanocomposite coatings during the CVD process. They can be summarized as follows [109–111]:

Pre-Interaction: Convection transfer of reactive materials to the metal or nonmetallic substrate by the formation of diffusive flow.

Step 1: Penetration of the precursor and gaseous products (desired coating) through the boundary layer to the metallic or nonmetallic substrate.

Step 2: Absorption of the reactants or gaseous products (coatings) on the metallic or nonmetallic substrate.

Step 3: Homogeneous and equilibrium bottom-up reactions in the gas phase with the main components of the gas phase.

Step 4: Surface diffusion and surface reactions at the interface of the substrate.

Step 5: Separation and removal of non-consumable byproducts from the substrate interface (reaction site).

Step 6: Diffusive transport of gaseous products separated from the surface by the boundary layer at the interface.

Step 7: Convective transfer of subproducts through the boundary layer by diffusive flow force.

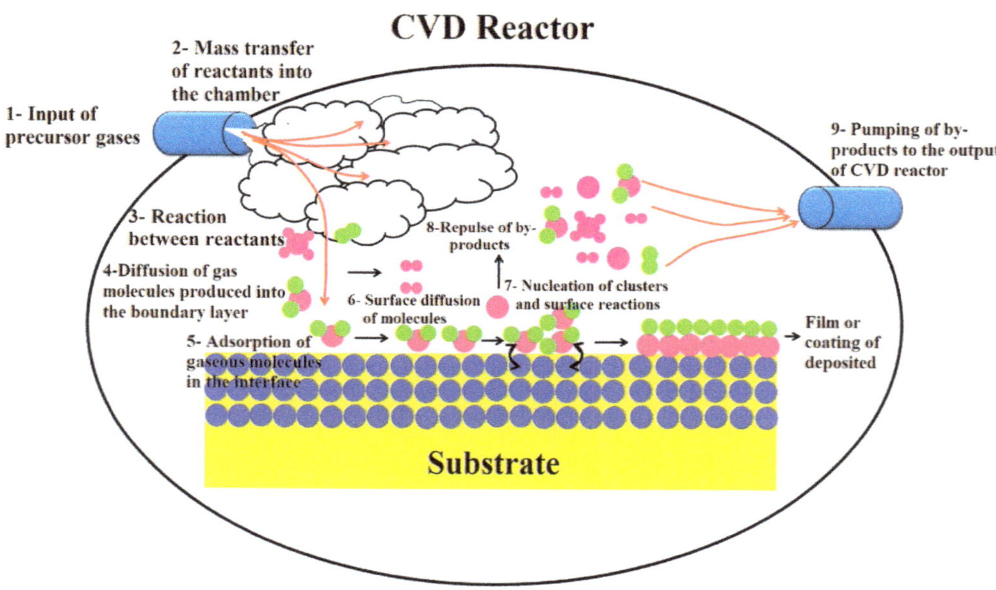

Figure 5. Different steps for deposition of ceramic nanocomposite coatings in the CVD technology.

The precondition for the synthesis of ceramic coatings by the CVD process is the presence of volatile compounds that include the compounds needed to precipitate these ceramic coatings. Although gas precursors can be used directly, the liquid and solid precursors must evaporate first before entering the reactor. Usually, liquids are heated in a cylinder, and solids are heated in a tubular evaporator and then evaporated at high temperatures. They are then atomized or become an aerosol. Chlorides, fluorides, and bromides are present in CVD technology, where they contain various metals in the liquid or gas state. These materials are volatile, and they are suitable alternatives for the deposition of different materials on various substrates [112–118].

Most of the classical processes are based on the law of chemical synthesis. However, this depends on which halogenated, compounds in the form of reductioning or oxidizing gas, move towards the substrate and precipitate on it. In particular, most of the halogenated compounds are reactive at temperatures above 750–900 °C. This point should be taken into account in the production of ceramic coating via CVD technology. Typically, chemical synthesis involves chemical reactions such as hydrogen reduction, reductioning, metal reduction, oxidation, hydrolysis, carbidization (for ceramic coatings), and nitrogenization. The most important advantage of using CVD technology in the coating industry is the selective deposition of different coating layers. In fact, using the CVD technology, thin films or coatings can be selectively deposited on various substrates. Layer selection is based on differences in chemical reactions occurring between the surface of a material and the reactants. This method can produce various nanostructures or nanoparticles such as ceramic nanostructures, carbides, carbon nanotubes, and even diamonds [114–118]. This is one of the strengths of CVD technology in the coatings industry. Due to the high synthesis speed of the process, it can be manufactured using different nanostructures. Additionally, due to the high reaction temperature of the CVD chamber, a variety of precursors can be employed. Another advantage of this method is the absence of large amounts of byproducts. Byproducts are generally gaseous and can easily be separated from the products in the compartment. Therefore, if pure raw materials or precursors are used in the process, pure products are also obtained. This is performed in such a way that the byproducts are negligible and easily removed from the environment. As a result, the primary product generated from the reaction of the precursor with the substrate is easily separated from the byproducts and the raw materials. If the goal is to produce oxide materials, the reaction can take place in the natural atmosphere, thus making the device simpler and cheaper. In sum, the composition of the desired products can be controlled by changing the type of substrate, the reaction, structure, shape, and [114–118]. Alteration of metallurgical conditions (environment, properties, and production process), often by the application of high-performance oxide coating layers, can solve problems such as corrosion, wear, conductivity, friction, etc. of the manufactured pieces [119–122].

5. Summary

The CVD technology is a process in which volatile compounds of different materials are used to deposit a nonvolatile solid on an appropriate substrate. This short review explains the industrial usage, various steps of the process, and the important factors influencing the products. Nowadays, CVD technology is widely used for the production of thin films and ceramic coatings. The CVD processes are commonly performed at high temperatures and low pressures. The three main steps of the technology are surface reaction (kinetic), diffusion (mass transfer), and desorption. The CVD process is divided into several steps according to the type of heat source, temperature, and pressure of the chamber. Each of these methods has advantages and disadvantages based on cost, uniformity of the coating, ability to create ceramic nanocomposite coatings, and characteristic control of the thin film that is produced. Moreover, important factors in terms of the properties of the coating deposited by CVD technology are the dimensions or geometry of the substrate, the shape of the substrate, the substrate temperature, the chemical composition of the substrate, the type of deposition process, the temperature inside the chamber, the purity of the target material, and the economic factors of the production. Finally, based on the review of this study, it is predicted that nanolayer ceramic and advanced ceramic coating industries would be the new venues for the applications of CVD technology to produce coatings usable at very high temperatures, particularly for hot corrosion purposes.

Author Contributions: M.S. (Masoud Sabzi): Writing—original draft; S.H.M.A.: Writing—original draft, Supervision; M.S. (Milad Shamsodin): Data curation; M.F.: Writing—review and editing; A.H.-N.: Writing—review and editing; P.F.: Writing—review and editing; N.P.: Writing—review and editing; U.L.: Writing—review and editing. All authors have read and agreed to the published version of the manuscript.

Funding: This research received no external funding.

Institutional Review Board Statement: Not applicable.

Informed Consent Statement: Not applicable.

Data Availability Statement: Not applicable.

Conflicts of Interest: The authors declare that they have no conflict of interest.

References

1. Hallad, S.A.; Banapurmath, N.R.; Hunashyal, A.M.; Shettar, A.S. Study of the effect of nanocomposite thin film coating on cutting tool tip for tribological applications. *Mater. Today Proc.* **2020**, *27*, 37–39. [CrossRef]
2. Sabzi, M.; Mousavi Anijdan, S.H.; Asadian, M. The effect of substrate temperature on microstructural evolution and hardenability of tungsten carbide coating in hot filament chemical vapor deposition. *Int. J. Appl. Ceram. Technol.* **2018**, *15*, 1350–1357. [CrossRef]
3. Perez-Mariano, J.; Lau K, H.; Alvarez, E.; Malhotra, R. Coatings for corrosion protection of porous substrates in gasifier components, Surf. *Coat. Technol.* **2008**, *202*, 2794–2800. [CrossRef]
4. Delfini, A.; Vricella, A.; Bueno Morles, R.; Pastore, R.; Micheli, D.; Gugliermetti, F.; Marchetti, M. CVD nano-coating of carbon composites for space materials atomic oxygen shielding. *Procedia Struct. Integr.* **2017**, *3*, 208–216. [CrossRef]
5. Li, Z.; Wang, Y.; Xiong, X.; Li, X.; Chen, Z.; Sun, W. Microstructure and growth behavior of Hf (Ta)C ceramic coating synthesized by low pressure chemical vapor deposition. *J. Alloys Compd.* **2017**, *705*, 79–88. [CrossRef]
6. Mousavi Anijdan, S.H.; Sabzi, M.; Asadian, M.; Jafarian, H.R. Effect of sub-layer temperature during HFCVD process on morphology and corrosion behavior of tungsten carbide coating. *Int. J. Appl. Ceram. Technol.* **2019**, *16*, 243–253. [CrossRef]
7. Strauss, H.W.; Chromik, R.R.; Hassani, S.; Klemberg-Sapieha, J.E. In situ tribology of nanocomposite Ti–Si–C–H coatings prepared by PE-CVD. *Wear* **2011**, *272*, 133–148. [CrossRef]
8. Feng, B.; Cao, D.M.; Meng, W.J.; Rehn, L.E.; Baldo, P.M.; Doll, G.L. Probing for mechanical and tribological anomalies in the TiC/amorphous hydrocarbon nanocomposite coating system. *Thin Solid Film.* **2001**, *398–399*, 210–216. [CrossRef]
9. Stueber, M.; Albers, U.; Leiste, H.; Ulrich, S.; Holleck, H.; Barna, P.B.; Kovacs, A.; Hovsepian, P.; Gee, I. Multifunctional nanolaminated PVD coatings in the system Ti–Al–N–C by combination of metastable fcc phases and nanocomposite microstructures. *Surf. Coat. Technol.* **2006**, *200*, 6162–6171. [CrossRef]
10. Tamilarasan, T.R.; Rajendran, R.; Rajagopal, G.; Sudagar, J. Effect of surfactants on the coating properties and corrosion behaviour of Ni–P–nano-TiO_2 coatings. *Surf. Coat. Technol.* **2015**, *276*, 320–326. [CrossRef]
11. Sabzi, M.; Mersagh Dezfuli, S. Deposition of Al_2O_3 ceramic film on copper-based heterostructured coatings by aluminizing process: Study of the electrochemical responses and corrosion mechanism of the coating. *Int. J. Appl. Ceram. Technol.* **2019**, *16*, 195–210. [CrossRef]
12. Shahriari, A.; Aghajani, H. Electrophoretic deposition of 3YSZ coating on AZ91D using an aluminum interlayer. *Prot. Met. Phys. Chem. Surf.* **2017**, *53*, 518–526. [CrossRef]
13. Wang, J.B.; Ren, Z.; Hou, Y.; Yan, X.L.; Liu, P.Z.; Zhang, H.; Zhang, H.X.; Guo, J.J. A review of graphene synthesis at low temperatures by CVD methods. *New Carbon Mater.* **2020**, *35*, 193–208. [CrossRef]
14. Pammi, S.V.N.; Maddaka, R.; Tran, V.-D.; Eom, J.-H.; Pecunia, V.; Majumder, S.; Kim, M.-D.; Yoon, S.G. CVD-deposited hybrid lead halide perovskite films for high-responsivity, self-powered photodetectors with enhanced photo stability under ambient conditions. *Nano Energy* **2020**, *74*, 104872. [CrossRef]
15. Shahriari, A.; Aghajani, H. Electrophoretic Deposition of 3YSZ Coating on AZ91D Alloy Using Al and Ni-P Interlayers. *J. Mater. Eng. Perform.* **2016**, *25*, 4369–4382. [CrossRef]
16. Pedersen, H.; Barry, S.T.; Sundqvist, J. Green CVD—Toward a sustainable philosophy for thin film deposition by chemical vapor deposition. *J. Vac. Sci. Technol. A* **2021**, *39*, 051001. [CrossRef]
17. Hong, Y.L.; Liu, Z.; Wang, L.; Zhou, T.; Ma, W.; Xu, C.; Feng, S.; Chen, L.; Chen, M.L.; Sun, D.M.; et al. Chemical vapor deposition of layered two-dimensional $MoSi_2N_4$ materials. *Science* **2020**, *369*, 670–674. [CrossRef]
18. Alagh, A.; Annanouch, F.E.; Umek, P.; Bittencourt, C.; Sierra-Castillo, A.; Haye, E.; FrançoisColomer, J.; Llobet, E. CVD growth of self-assembled 2D and 1D WS_2 nanomaterials for the ultrasensitive detection of NO_2. *Sens. Actuators B Chem.* **2021**, *326*, 128813. [CrossRef]
19. Tamilarasan, T.R.; Sanjith, U.; Rajendran, R.; Rajagopal, G.; Sudagar, J. Effect of Reduced Graphene Oxide Reinforcement on the Wear Characteristics of Electroless Ni-P Coatings. *J. Mater. Eng. Perform.* **2018**, *27*, 3044–3053. [CrossRef]
20. Tang, L.; Li, T.; Luo, Y.; Feng, S.; Cai, Z.; Zhang, H.; Liu, B.; Cheng, H.M. Vertical Chemical Vapor Deposition Growth of Highly Uniform 2D Transition Metal Dichalcogenides. *ACS Nano* **2020**, *14*, 4646–4653. [CrossRef]
21. Raiford, J.A.; Oyakhire, S.T.; Bent, S.F. Applications of atomic layer deposition and chemical vapor deposition for perovskite solar cells. *Energy Environ. Sci.* **2020**, *13*, 1997–2023. [CrossRef]
22. Paradisanos, I.; Shree, S.; George, A.; Leisgang, N.; Robert, C.; Watanabe, K.; Taniguchi, T.; Warburton, R.J.; Turchanin, A.; Marie, X.; et al. Controlling interlayer excitons in MoS2 layers grown by chemical vapor deposition. *Nat. Commun.* **2020**, *11*, 2391. [CrossRef] [PubMed]
23. Li, G.; Yin, S.; Tan, C.; Chen, L.; Yu, M.; Li, L.; Yan, F. Fast Photothermoelectric Response in CVD-Grown $PdSe_2$ Photodetectors with In-Plane Anisotropy. *Adv. Funct. Mater.* **2021**, *31*, 2104787. [CrossRef]
24. Zhang, J.; Wang, F.; Shenoy, V.B.; Tang, M.; Lou, J. Towards controlled synthesis of 2D crystals by chemical vapor deposition (CVD). *Mater. Today* **2020**, *40*, 132–139. [CrossRef]

25. Sabzi, M.; Mousavi Anijdan, S.H. Microstructural Analysis and Optical Properties Evaluation of Sol-Gel NiO-TiO_2 Heterostructured Films used for Solar Panels. *Ceram. Int.* **2019**, *45*, 3250–3255. [CrossRef]
26. Zhang, P.; Zhang, Y.; Chen, G.; Gai, W.; Kong, J. High-temperature oxidation behavior of CVD-SiC ceramic coating in wet oxygen and structural evolution of oxidation product: Experiment and first-principle calculations. *Appl. Surf. Sci.* **2021**, *556*, 149808. [CrossRef]
27. Tong, M.; Fu, Q.; Liang, M.; Feng, T.; Hou, W.; Sun, J.; Shi, X. Effect of PyC-SiC double-layer interface on ablation behaviour of impacted CVD-SiCnws/HfC coating. *Corr. Sci.* **2021**, *191*, 109741. [CrossRef]
28. Kong, J.; Zhang, Y.; Gai, W.; Wang, H.; Chen, G.; Zhang, P.; Li, H. Influence of crystallite morphology on the ablative behaviors of CVD-TaC coatings prepared on C/C composites beyond 2100 °C. *Corr. Sci.* **2022**, *205*, 110426. [CrossRef]
29. Zhang, J.; Zhang, Y.; Chen, R.; Zhu, X.; Fu, Y. Effect of microstructure on the ablation behavior and mechanical properties of CVD-HfC coating. *Corr. Sci.* **2021**, *192*, 109815. [CrossRef]
30. Baux, A.; Jacques, S.; Allemand, A.; Vignoles, G.L.; David, P.; Piquero, T.; Stempin, M.-P.; Chollon, G. Complex geometry macroporous SiC ceramics obtained by 3D-printing, polymer impregnation and pyrolysis (PIP) and chemical vapor deposition (CVD). *J. Eur. Ceram. Soc.* **2021**, *41*, 3274–3284. [CrossRef]
31. Honda, Y.; Mikami, Y.; Inoue, M.; Shinag, K.; Abe, T. Low-temperature SiO_2 film coatings onto Cu particles using the polygonal barrel-plasma chemical vapor deposition method. *Appl. Surf. Sci.* **2022**, *588*, 152646. [CrossRef]
32. Fu, J.; Sun, Y.; Ji, Y.; Zhang, J. Fabrication of robust ceramic based superhydrophobic coating on aluminum substrate via plasma electrolytic oxidation and chemical vapor deposition methods. *J. Mater. Process. Technol.* **2022**, *306*, 117641. [CrossRef]
33. Tong, M.; Fu, Q.; Hu, D.; Zhou, L.; Feng, T. Improvement of ablation resistance of CVD-HfC/SiC coating on hemisphere shaped C/C composites by introducing diffusion interface. *J. Eur. Ceram. Soc.* **2021**, *41*, 4067–4075. [CrossRef]
34. Mallick, R.; Kumar, R.; Panda, A.; Sahoo, A.K. Hard turning performance evaluation using CVD and PVD coated carbide tools: A comparative study. *Surf. Rev. Lett.* **2022**, *29*, 2250020. [CrossRef]
35. Zhang, J.; Zhang, Y.; Fu, Y.; Zhang, Y.; Zhu, X. Growth mechanism and ablation behavior of CVD-HfC coating on the surface of C/C composites and CVD-SiC coating. *Corr. Sci.* **2021**, *192*, 109819. [CrossRef]
36. Li, B.; Li, H.; Yao, X.; Zhu, X.; Liu, N. Preparation and ablation resistance of ZrC nanowires-reinforced CVD-ZrC coating on sharp leading edge C/C composites. *Appl. Surf. Sci.* **2022**, *584*, 152617. [CrossRef]
37. Wang, C. Fabrication of porous SiC nanostructured coatings on C/C composite by laser chemical vapor deposition for improving the thermal shock resistance. *Ceram. Int.* **2022**, *48*, 12450–12459. [CrossRef]
38. Zheng, X.; Liu, Y.; Cao, Y.; Wang, J.; Zhang, Y. CVD synthesis of nanometer SiC coating on diamond particles. *Ceram. Int.* **2021**, *47*, 16162–16169. [CrossRef]
39. Zhang, Y.; Yu, L.; Fu, T.; Wang, J.; Shen, F.; Cui, K. Microstructure evolution and growth mechanism of Si-MoSi2 composite coatings on TZM (Mo-0.5Ti-0.1Zr-0.02 C) alloy. *J. Alloys Compd.* **2022**, *894*, 162403. [CrossRef]
40. Guo, Z.; Guo, B.; Zhao, Q.; Liu, W.; Zheng, Q. Optimisation of spray-mist-assisted laser machining of micro-structures on CVD diamond coating surfaces. *Ceram. Int.* **2021**, *47*, 22108–22120. [CrossRef]
41. Pardo, A.; Gómez-Aleixandre, C.; Celis, J.P.; Buijnsters, J.G. Friction and wear behavior of plasma assisted chemical vapor deposited nanocomposites made of metal nanoparticles embedded in a hydrogenated amorphous carbon matrix. *Surf. Coat. Technol.* **2012**, *206*, 3116–3124. [CrossRef]
42. Tynyshtykbayev, K.B.; Ainabayev, A.; Kononenko, O.; Chichkov, M.; Ramazanova, Z.; Zulkharnai, R.; Roshchupkin, D.; Sorokupudov, J.; Starkov, V.; Insepov, Z. Low temperature synthesis of graphene nanocomposites using surface passivation of porous silicon nanocrystallites with carbon atoms. *Diam. Relat. Mater.* **2019**, *92*, 53–60. [CrossRef]
43. Perez-Mariano, J.; Caro, J.; Colominas, C. TiN/SiNx submicronic multilayer coatings obtained by chemical vapor deposition in a fluidized bed reactor at atmospheric pressure (AP/FBR-CVD). *Surf. Coat. Technol.* **2006**, *201*, 4021–4025. [CrossRef]
44. Bobzin, K. High-performance coatings for cutting tools. *CIRP J. Manuf. Sci. Technol.* **2017**, *18*, 1–9. [CrossRef]
45. Jang, J.; Kang, I.; Yi, K.W.; Cho, Y.W. Highly conducting fibrous carbon-coated silicon alloy anode for lithium ion batteries. *Appl. Surf. Sci.* **2018**, *454*, 277–283. [CrossRef]
46. Meng, W.J.; Zhang, X.D.; Shi, B.; Jiang, J.C.; Rehn, L.E.; Baldo, P.M.; Tittsworth, R.C. Structure and mechanical properties of Ti–Si–N ceramic nanocomposite coatings. *Surf. Coat. Technol.* **2003**, *163–164*, 251–259. [CrossRef]
47. Leem, M.; Lee, H.; Park, T.; Ahn, W.; Kim, H.; Lee, E.; Kim, H. Intriguing morphological evolution during chemical vapor deposition of HfS_2 using HfCl4 and S on sapphire substrate. *Appl. Surf. Sci.* **2020**, *509*, 144701. [CrossRef]
48. Hoyos-Palacio, L.M.; Paola Cuesta Castro, D.; Cristina Ortiz-Trujillo, I.; Elena Botero Palacio, L.; Janeth Galeano Upegui, B.; Javier Escobar Mora, N.; Antonio Carlos Cornelio, J. Compounds of carbon nanotubes decorated with silver nanoparticles via in-situ by chemical vapor deposition (CVD). *J. Mater. Res. Technol.* **2019**, *8*, 5893–5898. [CrossRef]
49. Sevgili, E.; Karaman, M. Initiated chemical vapor deposition of poly(Hydroxypropyl methacrylate) thin films. *Thin Solid Film.* **2019**, *687*, 137446. [CrossRef]
50. Méndez-Lozano, N.; Apátiga-Castro, M.; Manzano-Ramírez, A.; Rivera-Muñoz, E.M.; Velázquez-Castillo, R.; Alberto-González, C.; Zamora-Antuñano, M. Morphological study of TiO_2 thin films doped with cobalt by Metal Organic Chemical Vapor Deposition. *Results Phys.* **2020**, *16*, 1–5. [CrossRef]
51. Wang, M.; Jia, L.; Xu, H.; Li, A.; Peng, Y.; Tang, Z. Influence of pressure on chemical vapor deposition of boron nitride from $BCl_3/NH_3/H_2$ gas mixtures. *Ceram. Int.* **2020**, *46*, 4843–4849. [CrossRef]

52. Feng, X.; Jiang, L.; Li, D.; Tian, S.; Zhu, X.; Wang, H.; He, C.; Li, K. Progress and key challenges in catalytic combustion of lean methane. *J. Energy Chem.* **2022**, *75*, 173–215. [CrossRef]
53. Zhang, Z.; Yang, F.; Zhang, H.; Zhang, T.; Wang, H.; Xu, Y.; Ma, Q. Influence of CeO_2 addition on forming quality and microstructure of TiCx-reinforced $CrTi_4$-based laser cladding composite coating. *Mater. Charact.* **2021**, *171*, 110732. [CrossRef]
54. Morali, U.; Demiral, H.; Sensoz, S. Synthesis of carbon molecular sieve for carbon dioxide adsorption: Chemical vapor deposition combined with Taguchi design of experiment method. *Powder Technol.* **2019**, *355*, 716–726. [CrossRef]
55. Zhao, Y.; Song, B.; Cui, X.; Ren, Y.; Yu, W.; Wang, Y. High electrocatalytic reduction using ZnS micropolyhedron: Direct sulfuration of ZIF-8 film on conducting glass by chemical vapour deposition. *Mater. Lett.* **2019**, *250*, 193–196. [CrossRef]
56. Lee, S.; Park, W.K.; Yoon, Y.; Baek, B.; Yoo, J.S.; Kwon, S.B.; Kim, D.H.; Joon, Y.; Bong, H.; Kang, K.; et al. Quality improvement of fast-synthesized graphene films by rapid thermal chemical vapor deposition for mass production. *Mater. Sci. Eng. B* **2019**, *242*, 63–68. [CrossRef]
57. Matsumura, H. Current Status of Catalytic Chemical Vapor Deposition Technology—History of Research and Current Status of Industrial Implementation. *Thin Solid Film.* **2019**, *679*, 42–48. [CrossRef]
58. Zhang, S.; Wang, J.; Lu, Y.; Hou, W.; Cao, K.; Guo, S.; Wang, Z.; Wang, L. Highly ordered arrays and characterization of WS2 flakes grown by low pressure chemical vapour deposition. *Chem. Phys.* **2019**, *523*, 106–109. [CrossRef]
59. Zou, C.; Li, B.; Liu, K.; Yang, X.; Li, D. Microstructure and mechanical properties of Si3N4f/BN composites with BN interphase prepared by chemical vapor deposition of borazine. *J. Eur. Ceram. Soc.* **2020**, *40*, 1139–1148. [CrossRef]
60. Katamune, Y.; Arikaw, D.; Mori, D.; Izumi, A. Formation of phosphorus-incorporated diamond films by hot-filament chemical vapor deposition using organic phosphorus solutions. *Thin Solid Film.* **2019**, *677*, 28–32. [CrossRef]
61. Madaka, R.; Kumari, J.; Kanneboina, V.; Jha, S.H.; Agarwal, P. Role of chamber pressure on crystallinity and composition of silicon films using silane and methane as precursors in hot-wire chemical vapour deposition technique. *Thin Solid Film.* **2019**, *682*, 126–130. [CrossRef]
62. Liu, Q.; Liu, J.; Luan, X. Preparation of ZrC-SiC composite coatings by chemical vapor deposition and study of co-deposition mechanism. *J. Mater. Sci. Technol.* **2019**, *35*, 2942–2949. [CrossRef]
63. Tarazona, A.; Bucio, T.D.; Oo, S.Z.; Petra, R.; Khokhar, A.Z.; Boden, S.A.; Gardes, F.Y.; Reed, G.T.; Chong, H.M. Hot wire chemical vapor deposition for silicon photonics: An emerging industrial application opportunity. *Thin Solid Film.* **2019**, *676*, 26–30. [CrossRef]
64. Luo, J.; Wang, J.; Xia, F.; Huang, X. Direct growth of large area uniform double layer graphene films on MgO(100) substrates by chemical vapor deposition. *Mater. Chem. Phys.* **2019**, *233*, 213–219. [CrossRef]
65. Chuang, C.; Mineharu, M.; Matsunag, M.; Liu, C.W.; Wu, B.Y.; Kenj, J.H.; Taniguchi, T.; Liang, C.T.; Aoki, N. Conductance interference effects in an electron-beam-resist-free chemical vapor deposition graphene device sandwiched between two h-BN sheets. *Carbon* **2019**, *154*, 238–243. [CrossRef]
66. Basak, S.; Afzaal, M.; Yates, H.M. Optically tuned and large-grained bromine doped $CH_3NH_3PbI_3$ perovskite thin films via aerosol-assisted chemical vapour deposition. *Mater. Chem. Phys.* **2019**, *223*, 157–163. [CrossRef]
67. Silverio, V.; Canane, P.A.G.; Cardoso, S. Surface wettability and stability of chemically modified silicon, glass and polymeric surfaces via room temperature chemical vapor deposition. *Colloids Surf. A Physicochem. Eng. Asp.* **2019**, *570*, 210–217. [CrossRef]
68. Dong, H.; Duan, A.; Zhong, S.; Zhang, Y.; Zhou, X. Chemical vapor deposition growth of broadband tunable light absorption and anti-reflection properties of VS2 nanostructure films. *Mater. Lett.* **2019**, *252*, 227–230. [CrossRef]
69. Huang, Y. Modeling and analysis of SiO2 deposition during high-purity fused silica glass synthesis by $SiCl_4$ chemical vapor deposition. *Ceram. Int.* **2019**, *45*, 10740–10745. [CrossRef]
70. Wang, Z.; Yue, H.Y.; Yu, Z.M.; Huang, Z.; Gao, X.; Wang, B.; Song, S.S.; Guan, E.H.; Wang, W.Q.; Zhang, H.J. A novel 3D porous graphene foam prepared by chemical vapor deposition using nickel nanoparticles: Electrochemical determination of levodopa in the presence of uric acid. *Microchem. J.* **2019**, *147*, 163–169. [CrossRef]
71. Liu, X.F.; Yan, G.G.; Shen, Z.W.; Wen, Z.X.; Chen, J.; He, Y.W.; Zhao, W.S.; Wang, L.; Guan, M.; Zhang, F. Homoepitaxial growth of multiple 4H-SiC wafers assembled in a simple holder via conventional chemical vapor deposition. *J. Cryst. Growth* **2019**, *507*, 283–287. [CrossRef]
72. Gutiérrez-García, C.J.; Madaí Ambriz-Torres, J.; Jesús Contreras-Navarrete, J.; Gabriel Granados-Martínez, F.; Litzajaya García-Ruiz, D.; García-González, L.; Zamora-Peredo, L.; Fernando Ortega-Varela, L.; Richaud, A.; Méndez, F.; et al. Synthesis of carbon spheres by atmospheric pressure chemical vapor deposition from a serial of aromatic hydrocarbon precursors. *Phys. E Low-Dimens. Syst. Nanostruct.* **2019**, *112*, 78–85. [CrossRef]
73. Arzaee, N.A.; Noh, M.F.; Aadenan, A.; Mumthas, I.N.; Ab Hamid, F.F.; Kamarudin, N.N.; Mohamed, N.A.; Ibrahim, M.A.; Ismail, A.F.; Teridi, M.A. Aerosol-assisted chemical vapour deposition of α-Fe_2O_3 nanoflowers for photoelectrochemical water splitting. *Ceram. Int.* **2019**, *45*, 16797–16802. [CrossRef]
74. Ning, B.; Xia, T.; Tong, Z.X.; He, Y.L. Experimental and numerical studies of tungsten line growth in laser chemical vapor deposition. *Int. J. Heat Mass Transf.* **2019**, *140*, 564–578. [CrossRef]
75. Singh, M.; Jha, H.S.; Agarwal, P. Synthesis of vertically aligned carbon nanoflakes by hot-wire chemical vapor deposition: Influence of process pressure and different substrates. *Thin Solid Film.* **2019**, *678*, 26–31. [CrossRef]
76. Sharma, A.; Patwardhan, A.; Dasgupt, K.; Joshi, J.B. Kinetic study of boron doped carbon nanotubes synthesized using chemical vapour deposition. *Chem. Eng. Sci.* **2019**, *207*, 1341–1352. [CrossRef]

77. Fan, X.; Sun, J.; Guo, W.; Ke, X.; Yan, C.; Li, X.; Dong, Y.; Xiong, F.; Fu, Y.; Wang, L.; et al. Chemical vapor deposition of graphene on refractory metals: The attempt of growth at much higher temperature. *Synth. Met.* **2019**, *247*, 233–239. [CrossRef]
78. Jia, K.; Ci, H.; Zhang, J.; Sun, Z.; Ma, Z.; Zhu, Y.; Liu, S.; Liu, J.; Sun, L.; Liu, X.; et al. Superclean Growth of Graphene Using a Cold-Wall Chemical Vapor Deposition Approach. *Angew. Chem.* **2020**, *59*, 17214–17218. [CrossRef]
79. Xu, K.; Liu, H.; Shi, Y.C.; You, J.Y.; Ma, X.Y.; Cui, H.J.; Yan, Q.B.; Chen, G.C.; Su, G. Preparation of T-carbon by plasma enhanced chemical vapor deposition. *Carbon* **2020**, *157*, 270–276. [CrossRef]
80. Ponja, S.D.; Sathasivam, S.; Parkin, I.P.; Carmalt, C.J. Highly conductive and transparent gallium doped zinc oxide thin films via chemical vapor deposition. *Sci. Rep.* **2020**, *10*, 638. [CrossRef]
81. Fraga, M.; Pessoa, R. Progresses in Synthesis and Application of SiC Films: From CVD to ALD and from MEMS to NEMS. *Micromachines* **2020**, *11*, 799. [CrossRef]
82. Yang, X.; Li, W.; Lv, J.; Sun, G.; Shi, Z.; Su, Y.; Lian, X.; Shao, Y.; Zhi, A.; Tian, X.; et al. In situ separator modification via CVD-derived N-doped carbon for highly reversible Zn metal anodes. *Nano Res.* **2021**, *15*, 9785–9791. [CrossRef]
83. Mwafy, E.A. Eco-friendly approach for the synthesis of MWCNTs from waste tires via chemical vapor deposition, Environmental Nanotechnology. *Monit. Manag.* **2020**, *14*, 100342.
84. Mersagh Dezfuli, S.; Sabzi, M. Deposition of ceramic nanocomposite coatings by electroplating process: A review of layer-deposition mechanisms and effective parameters on the formation of the coating. *Ceram. Int.* **2019**, *45*, 21835–21842. [CrossRef]
85. Mersagh Dezfuli, S.; Sabzi, M. Deposition of self-healing thin films by the sol–gel method: A review of layer-deposition mechanisms and activation of self-healing mechanisms. *Appl. Phys. A* **2019**, *125*, 1–8. [CrossRef]
86. Dai, C.; Li, B.; Li, J.; Zhao, B.; Wu, R.; Ma, H.; Duan, X. Controllable synthesis of NiS and NiS_2 nanoplates by chemical vapor deposition. *Nano Res.* **2020**, *13*, 2506–2511. [CrossRef]
87. Xia, M.; Zhou, Z.; Su, Y.; Li, Y.; Wu, Y.; Zhou, N.; Zhang, H.; Xiong, X. Scalable synthesis SiO@C anode by fluidization thermal chemical vapor deposition in fluidized bed reactor for high-energy lithium-ion battery. *Appl. Surf. Sci.* **2019**, *467–468*, 298–308. [CrossRef]
88. Saini, S.; Mele, P.; Oyake, T.; Shiomi, J.; Niemelä, J.P.; Karppinen, M.; Miyazaki, K.; Li, C.; Kawaharamura, T.; Ichinose, A.; et al. Porosity-tuned thermal conductivity in thermoelectric Al-doped ZnO thin films grown by mist-chemical vapor deposition. *Thin Solid Film.* **2019**, *685*, 180–185. [CrossRef]
89. Yin, T.; Jiang, B.; Su, Z.; Fan, Z.; Huang, Q. Numerical simulation of carrier gas effects on flow field, species concentration and deposition rate in the chemical vapor deposition of carbon. *New Carbon Mater.* **2018**, *33*, 357–363. [CrossRef]
90. Wang, X.; Wu, S.; Zhong, Y.; Wang, Y.; Pan, Y.; Tang, H. Electrochemically mediated decarboxylative acylation of N-nitrosoanilines with α-oxocarboxylic acids. *Chin. Chem. Lett.* **2022**, *34*, 107537. [CrossRef]
91. Ou, C.; Pan, Y.; Tang, H. Electrochemically promoted N-heterocyclic carbene polymer-catalyzed cycloaddition of aldehyde with isocyanide acetate. *Sci. China Chem.* **2022**, *65*, 1873–1878. [CrossRef]
92. An, L.; Zhang, T.; Lei, X.; Yang, P.; Liu, Y. Local heat and mass transfer characteristics of different channel configurations in polysilicon chemical vapor deposition reactor. *Sol. Energy* **2020**, *196*, 494–504. [CrossRef]
93. Akamatsu, K.; Suzuki, M.; Nakao, A.; Nakao, S. Development of hydrogen-selective dimethoxydimethylsilane-derived silica membranes with thin active separation layer by chemical vapor deposition. *J. Membr. Sci.* **2019**, *580*, 268–274. [CrossRef]
94. Feng, C.; Liu, X.; Wen, S.; An, Y. Controlled growth and characterization of In_2O_3 nanowires by chemical vapor deposition. *Vacuum* **2019**, *161*, 328–332. [CrossRef]
95. Yoshimune, M.; Haraya, K. Simple control of the pore structures and gas separation performances of carbon hollow fiber membranes by chemical vapor deposition of propylene. *Sep. Purif. Technol.* **2019**, *223*, 162–167. [CrossRef]
96. Huang, X.; Sun, S.; Lu, S.; Li, K.; Tu, G.; Song, J. Synthesis and characterization of oxidation-resistant TiB_2 coating on molybdenum substrate by chemical vapor deposition. *Mater. Lett.* **2018**, *228*, 53–56. [CrossRef]
97. Xin, B.; Sun, G.; Lao, C.; Shang, D.; Zhang, X.; Wen, Z.; He, M. Chemical vapor deposition synthesis of carbon nanosprouts on calcined stainless steel. *Mater. Lett.* **2019**, *238*, 290–293. [CrossRef]
98. Ciprian, M.; Xu, P.; Chaemchuen, S.; Tu, R.; Zhuiykov, S.; Heynderickx, P.M.; Verpoort, F. MoO_3 nanoparticle formation on zeolitic imidazolate framework-8 by rotary chemical vapor deposition. *Microporous Mesoporous Mater.* **2018**, *267*, 185–191. [CrossRef]
99. Zhao, X.; Gou, L. Comparative analysis of graphene grown on copper and nickel sheet by microwave plasma chemical vapor deposition. *Vacuum* **2018**, *153*, 48–52. [CrossRef]
100. He, C.; Bu, X.; Yang, S.; Hea, P.; Ding, G.; Xie, X. Core-shell $SrTiO_3$/graphene structure by chemical vapor deposition for enhanced photocatalytic performance. *Appl. Surf. Sci.* **2018**, *436*, 373–381. [CrossRef]
101. Bigiani, L.; Maccato, C.; Gasparotto, A.; Sada, C.; Barreca, D. Structure and properties of Mn3O4 thin films grown on single crystal substrates by chemical vapor deposition. *Mater. Chem. Phys.* **2019**, *223*, 591–596. [CrossRef]
102. Matsumoto, S.; Kaneda, Y.; Ito, A. Highly self-oriented growth of (020) and (002) monoclinic HfO_2 thick films using laser chemical vapor deposition. *Ceram. Int.* **2020**, *46*, 1810–1815. [CrossRef]
103. Khanna, S.; Utsav, B.; Patel, R.; Marathey, P.; Chaudari, R.; Vora, J.; Banerjee, R.; Ray, R.; Mukhopadhyay, I. Growth of titanium dioxide nanorod over shape memory material using chemical vapor deposition for energy conversion application. *Mater. Today Proc.* **2020**, *28*, 475–479. [CrossRef]
104. Katsui, H.; Kondo, N. Preferred orientations and microstructures of lanthanum phosphate films prepared via laser chemical vapor deposition. *J. Cryst. Growth* **2019**, *519*, 46–53. [CrossRef]

105. Ito, A.; Sekiyam, M.; Hara, T.; Goto, T. Self-oriented growth of β-$Yb_2Si_2O_7$ and X1/X2-Yb_2SiO_5 coatings using laser chemical vapor deposition. *Ceram. Int.* **2020**, *46*, 9548–9553. [CrossRef]
106. Liang, Y. Chemical vapor deposition synthesis of Ge doped ZnO nanowires and the optical property investigation. *Phys. Lett. A* **2019**, *383*, 2928–2932. [CrossRef]
107. Feng, X.; Xia, L.; Jiang, Z.; Tian, M.; Zhang, S.; He, C. Dramatically promoted toluene destruction over Mn@Na-Al_2O_3@Al monolithic catalysts by Ce incorporation: Oxygen vacancy construction and reaction mechanism. *Fuel* **2022**, *326*, 125051. [CrossRef]
108. Pan, Y.; Xia, Q.; Zhu, J.; Wang, Y.; Liang, Y.; Wang, H.; Tang, H.; Pan, Y. Electrochemically Mediated Carboxylative Cyclization of Allylic/Homoallylic Amines with CO_2 at Ambient Pressure. *Org. Lett.* **2022**, *24*, 8239–8243. [CrossRef]
109. Tkadletz, M.; Hofer, C.; Wüstefeld, C.; Schalk, N.; Motylenko, M.; Rafaj, D.; Holzschuh, H.; Bürgin, W.; Sartory, B.; Mitterer, C.; et al. Thermal stability of nanolamellar fcc-$Ti_{1-x}Al_xN$ grown by chemical vapor deposition. *Acta Mater.* **2019**, *174*, 195–205. [CrossRef]
110. Meng, Y.; Yin, H.; Liu, M.; Ma, T.; Jiang, S. Experimental study on the generation of carbonaceous dust formed by chemical vapor deposition in HTGR. *Nucl. Eng. Des.* **2018**, *335*, 172–177. [CrossRef]
111. Elghoul, N.; Kraiem, S.; Rahmouni, H.; Khirouni, K. Annealing effects on physical properties of a Au/a-Si:H Schottky diode prepared via the plasma-enhanced chemical vapor deposition technique. *J. Phys. Chem. Solids* **2019**, *132*, 18–25. [CrossRef]
112. Young Jang, D.; Deok Han, G.; Rak Choi, H.; Sik Kim, M.; Jong Choi, H.; Hyung Shim, J. $La_{0.6}Sr_{0.4}Co_{0.2}Fe_{0.8}O_3$-δ cathode surface-treated with La_2NiO_4+δ by aerosol-assisted chemical vapor deposition for high performance solid oxide fuel cells. *Ceram. Int.* **2019**, *45*, 12366–12371. [CrossRef]
113. Li, X.; Rafie, A.; Smolin, Y.Y.; Simotwo, S.; Kalra, V.; Lau, K.K.S. Engineering conformal nanoporous polyaniline via oxidative chemical vapor deposition and its potential application in supercapacitors. *Chem. Eng. Sci.* **2019**, *194*, 156–164. [CrossRef]
114. Demiral, H.; Demiral, I. Preparation and characterization of carbon molecular sieves from chestnut shell by chemical vapor deposition. *Adv. Powder Technol.* **2018**, *29*, 3033–3039. [CrossRef]
115. Najafabadi, A.H.; Mozaffarinia, R.; Rahimi, H.; Razavi, R.S.; Paimozd, E. Sol–gel processing of hybrid nanocomposite protective coatings using experimental design. *Prog. Org. Coat.* **2013**, *76*, 293–301. [CrossRef]
116. Hojjati Najafabadi, A.; Shoja Razavi, R.; Mozaffarinia, R.; Rahimi, H. A new approach of improving rain erosion resistance of nanocomposite sol-gel coatings by optimization process factors. *Metall. Mater. Trans. A* **2014**, *45*, 2522–2531. [CrossRef]
117. Qiu, L.; Du, Y.; Wang, S.; Du, L.; Chen, Z.; Wang, J.; Zhong, Z.; Shi, Z.; Albir, L. Through-process modeling and experimental verification of titanium carbonitride coating prepared by moderate temperature chemical vapor deposition. *Surf. Coat. Technol.* **2019**, *359*, 278–288. [CrossRef]
118. Schüler, C.; Betzenbichler, F.; Drescher, C.; Hinrichsen, O. Optimization of the synthesis of Ni catalysts via chemical vapor deposition by response surface methodology. *Chem. Eng. Res. Des.* **2018**, *132*, 303–312. [CrossRef]
119. Wu, R.; Tan, Y.; Meng, F.; Zhang, Y.; Huang, Y. PVDF/MAF-4 composite membrane for high flux and scaling-resistant membrane distillation. *Desalination* **2022**, *540*, 116013. [CrossRef]
120. Cheng, M.; Yan, X.; Cui, Y.; Han, M.; Wang, X.; Wang, J.; Zhang, R. An eco-friendly film of pH-responsive indicators for smart packaging. *J. Food Eng.* **2022**, *321*, 110943. [CrossRef]
121. Rahimi, H.; Mozafarinia, R.; Shoja Razavi, R.; Paimozd, E.; Hojjati-Najafabadi, A. *Processing and Properties of GPTMS-TEOS Hybrid Coatings on 5083 Aluminium Alloy, Advanced Materials Research*; Trans Tech Publications Ltd.: Bäch SZ, Switzerland, 2011; Volume 239.
122. Mahmoodi, S.; Hassan, D.A.; Hojjati-Najafabadi, A.; Li, W.; Liao, L.; Moshayedi, A.J.; Huang, X.; Khajavi, M.N. Quality enhancement of copper oxide thin film synthesized under elevated gravity acceleration by two-axis spin coating. *Ceram. Int.* **2020**, *46*, 7421–7429. [CrossRef]

coatings

MDPI

Article

Controlled Electroplating of Noble Metals on III-V Semiconductor Nanotemplates Fabricated by Anodic Etching of Bulk Substrates

Elena I. Monaico [1], Eduard V. Monaico [1,*], Veaceslav V. Ursaki [1,2] and Ion M. Tiginyanu [1,2]

1 National Center for Materials Study and Testing, Technical University of Moldova, 2004 Chisinau, Moldova
2 Academy of Sciences of Moldova, 2001 Chisinau, Moldova
* Correspondence: eduard.monaico@cnstm.utm.md

Abstract: Porous templates are widely used for the preparation of various metallic nanostructures. Semiconductor templates have the advantage of controlled electrical conductivity. Site-selective deposition of noble metal formations, such as Pt and Au nanodots and nanotubes, was demonstrated in this paper for porous InP templates prepared by the anodization of InP wafers. Metal deposition was performed by pulsed electroplating. The produced hybrid nanomaterials were characterized by scanning electron microscopy (SEM) and energy dispersive X-ray analysis (EDX). It was shown that uniform deposition of the metal along the pore length could be obtained with optimized pulse parameters. The obtained results are discussed in terms of the optimum conditions for effective electrolyte refreshing and avoiding its depletion in pores during the electroplating process. It was demonstrated that the proposed technology could also be applied for the preparation of metal nanostructures on porous oxide templates, when it is combined with thermal treatment for the oxidation of the porous semiconductor skeleton.

Keywords: pulsed electrodeposition; nanotubes; nanodots; porous template; varicap device; site-selective deposition

Citation: Monaico, E.I.; Monaico, E.V.; Ursaki, V.V.; Tiginyanu, I.M. Controlled Electroplating of Noble Metals on III-V Semiconductor Nanotemplates Fabricated by Anodic Etching of Bulk Substrates. *Coatings* **2022**, *12*, 1521. https://doi.org/10.3390/coatings12101521

Academic Editors: Konstantin Lovchinov and Gergana Alexieva

Received: 25 September 2022
Accepted: 7 October 2022
Published: 11 October 2022

1. Introduction

Over recent decades, the synthesis of functional nanomaterials has been a topic of interest due to their wide area of potential application in electronics, optoelectronics, photonics, magnetic devices, thermal energy storage, thermal transport, electrochromic windows, photovoltaic cells, biotechnology and life sciences, metallic building blocks in the form of nanowires, nanotubes and arrangements of nanoparticles, among others [1–7]. One of the main methods for the preparation of such structures is template synthesis, template electrodeposition being one of the most general approaches for the control of size, shape and structural properties of metallic nanostructures [8–10].

Apart from growing polycrystalline nanowires and nanotubes, single crystal metallic nanostructures can be achieved by controlling the electroplating parameters [8], electrodeposition being a cost-effective technique. Concerning the templates, etched ion track membranes based either on inorganic materials or on organic polymers [11,12], and porous alumina templates produced by anodization of aluminum foils [9,10,13–15], are the most widely used methods for nanofabrication purposes. Porous alumina templates are of special interest, because they can be produced with periodic arrangement of pores, thus, allowing the fabrication of periodically ordered metallic nanostructures, which enhances their use in device applications. High electrical resistivity is an essential drawback of porous alumina templates, which require additional functionalization of the inner surface of pores before electroplating metals, in order to produce metallic nanotubes. This drawback is overcome in semiconductor porous templates, due to the possibilities of controlling their conductivity. Porous templates have been demonstrated on Si, GaAs, GaP, InP, ZnSe and other semiconductors, including those with an ordered arrangement of pores or with pores of specific shape [16,17]. Nevertheless, uniform deposition of metals inside porous templates remains a challenge. Pulsed electrodeposition has been proposed instead of direct current (DC) electroplating to address this issue [17–21].

The comparative study of continuous current and pulsed electrodeposition of metals on flat substrates revealed some advantages of pulsed electrodeposition, related to superior quality due to reduced contamination by impurities and a lower rate of occluded gases [22,23], as well as smaller grain size at the long pause between pulses due to the larger number of nucleation sites [22,23]. Pulsed electrodeposition is even more promising for porous templates, especially those with pore diameters below 100 nm; since uniform deposition inside pores with continuous current deposition is hardly achievable, a thorough optimization of current pulse parameters being necessary to reach this goal [17,21].

Apart from the templated growth of various metallic nanostructures, randomly oriented porous metallic formations (for instance, nanoporous gold) can be produced by localized electrochemical dealloying [24]. While ordered templated growth is more often applied in electronics, optoelectronics and photonics, dealloying can be applied in high-strength material, catalysis, energy storage or biosensors. Atomically smooth starting surfaces prepared by sputter-annealing cycles in ultrahigh vacuum are usually needed to control the process of dealloying, while the electrochemical preparation of semiconductor porous templates for metal electrodeposition is less sensitive to the surface state, since the pore nucleation layer is removed after anodization. Moreover, while dealloying strongly depends on the crystallographic substrate orientation, the crystallographic orientation of the semiconductor substrate is important only for the growth of crysto pores, while the growth of current line-oriented pores is independent of the crystallographic orientation of the substrate [17].

Recently, an enhancement of the photocatalytic performances of aero-Ga_2O_3 induced by functionalization with noble metals, was demonstrated [25]. Due to the fact that gallium oxide exhibits high electrical resistivity, functionalization with noble metals via electrochemical deposition is challenging. In their study, the authors used an approach based on a template composed of a network of ZnO microtetrapods. The technological process consisted of four technological steps. In the first step, sputtering of gold was achieved on commercially available ZnO microtetrapods. In the second stage, a film of GaN was grown by hydride vapor phase epitaxy (HVPE) on Au-functionalized ZnO microtetrapods. In the process of HVPE growth of GaN, a simultaneous etching of ZnO occurred, which resulted in the formation of hollow GaN microtubes with the inner surface decorated by Au. At the third step, functionalization with Au of the outer surface of the GaN microtubes was obtained via sputtering. At the end, the GaN/Au microtubes were subjected to thermal annealing for transformation into aero-Ga_2O_3/Au. However, it should be noted that uniform sputtering in the depth of the template is difficult to achieve. Apart from that, this technological process is too complex. In this paper, we propose a feasible cost-effective approach allowing uniform functionalization of an array of Ga_2O_3 nanowires with Au nanodots. The technological process consists of electrochemical deposition of Au nanodots on a template of GaAs nanowires produced by anodization of bulk GaAs substrates, followed by oxidation of GaAs nanowires via their thermal annealing.

Pt and Au decorated templates could be promising for the development of microengines or micro-submarines with cargo capabilities and enhanced photocatalytic performances [26].

The goal of this paper is to demonstrate controlled electrodeposition of metals into different parts of porous InP templates by adjusting the parameters of the voltage pulses.

2. Materials and Methods

2.1. Anodization

Crystalline 500-μm thick *n*-InP(100) and *n*-GaP(100) substrates with a free electron concentration of 1.3×10^{18} cm^{-3} and 2×10^{17} cm^{-3}, respectively, were supplied by CrysTec

GmbH, Germany. Si-doped (111)B-oriented n-type GaAs wafers with a free electron concentration of 2×10^{18} cm^{-3} were acquired from Mateck GmbH, Jülich, Germany. The samples, with an electrical contact of a silver paste, were pressed against an O-ring in a Teflon cell, as shown in Figure 1. The electrochemical etching was carried out at room temperature in 3.5M NaCl and 1M HBr electrolytes at applied potentials (U) of 6 V and 15 V, respectively, in three electrode configurations, with a Pt mesh with the surface area of 6 cm^2 acting as a counter electrode. A saturated Ag/AgCl reference electrode, and the sample as working electrode, were used. For the purpose of obtaining GaAs nanowires, the anodization was conducted on Si-doped (111)B-oriented n-type GaAs wafers in 1M HNO_3 electrolyte at an applied potential of 4 V for 20 min. The potentiostat ELIPOR-2 (ET&TE Etch and Technology GmbH, Kiel, Germany) was fully controlled via a PC unit. The steering of the electrolyte was provided by a Teflon agitator connected to a motor. The rotation speed of 100 r·min^{-1} was controlled by the applied potential from an external multichannel power supply McPower LAB-2305 (ETT Marketing GmbH, Braunschweig, Germany) to the motor.

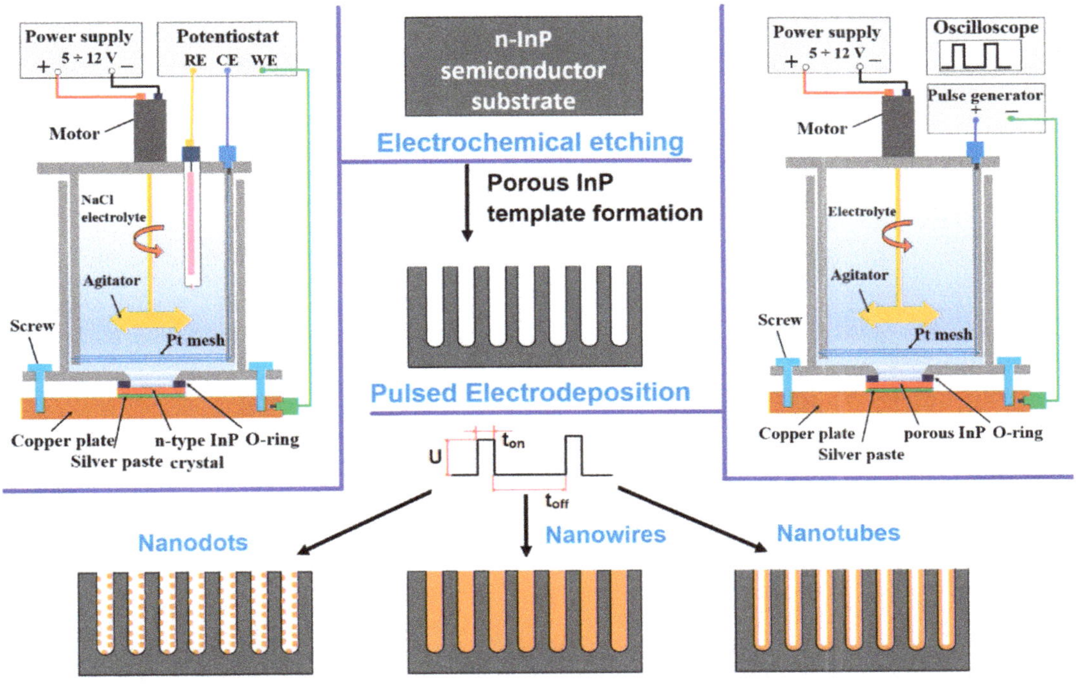

Figure 1. Schematic representation of the set-up for electrochemical etching of semiconductor substrates as well as pulsed electrochemical deposition of metal nanostructures in porous semiconductor templates. The technological work-flow is presented for InP crystals.

The morphology and chemical composition analyses of the porous semiconductor templates were investigated using Zeiss (Jena, Germany) Sigma and Tescan (Brno, Czech Republic) Vega TS 5130 MM scanning electron microscopes (SEM), equipped with an Oxford Instruments (Oxford, UK) INCA Energy EDX system operated at 20 kV.

2.2. Electroplating

For opening the entrance in pores after porosification, the top nucleation layer of porous InP samples was removed using isotropic wet etching by immersing of the specimens in $HCl:H_3PO_4$ electrolyte with a ratio of 1:1 for 25 s (T = 23 °C). The removal of the top irregular porous layer in GaP was conducted in a mixture of $HCl:HNO_3:H_2O$ (1:2:2) at 60 °C for 10 min in a well-ventilated fume-hood. The samples were immersed in the deposition electrolyte for 3 min prior to electrodeposition with the purpose of allowing the penetration of the metal ions inside the pores. Pulsed electrochemical deposition of Au and Pt was obtained in a commercially available gold bath containing 5 $g \cdot L^{-1}$ Au or 5 $g \cdot L^{-1}$ Pt (DODUCO GmbH, Pforzheim, Germany). The electrochemical deposition was performed at a temperature of 25 °C in a common two-electrode plating cell where the porous sample served as a working electrode, while a platinum wire was used as a counter electrode (Figure 1). A home-made generator capable of providing rectangular pulses with a pulse width (t_{on}) from 500 ns up to 10 s, and delay between pulses (t_{off}) up to 10 s, was used for electroplating. Metal nanodots, nanowires or nanotubes were deposited depending on the applied pulse width, voltage pulse amplitude (U), and electroplating duration, as illustrated in Figure 1. After electroplating, the samples were rinsed in distilled water.

3. Results and Discussion

The behavior of the electrochemical dissolution of semiconductor compounds governed by the breakdown mechanism, as well the type of pores obtained as a result of anodization, were reported in our review paper [17]. Briefly, the pore formation mechanism can be explained as follows: by applying the positive potential to the semiconductor, the holes start to drift towards the semiconductor–electrolyte interface (SEI). For n-type semiconductors, the holes are minority carriers. As a result, the anodic current remains low. With a further increase in the applied potential, at a certain potential, depending on the doping and defect density, an increase in current is usually observed in the current–voltage (I-V) characteristic [21]. This increase is related to the avalanche process in the space charge region (SCR). Usually, the anodic dissolution starts at surface defects (dislocations) marked as the root nucleus in Figure 2a. With further increase in the applied voltage, new electron–hole pairs will be generated as a result of multiple collisions of electrons tunneling through the SCR with the atoms. Once new electro–hole pairs are generated due to the breakdown, the etching around the defect will develop through branching of crystallographically oriented pores. At the next stage of anodization, the branching of pores stops, and the current line-oriented pores start to grow in perpendicular alignment to the substrate surface. It should be mentioned that current line-oriented pores cannot intersect each other, therefore, conditions are provided for the self-ordering of pores [17].

Electroplating of Pt on porous InP samples immediately after the electrochemical etching resulted in non-uniform deposition of nanotubes due to the top nucleation layer, as can be seen in Figure 2b. After a certain time, the pores in the nucleation layer were blocked with the deposited metal, which prevented the flow of the electrolyte into the full depth of the pores. Usually, the thickness of the nucleation layer in samples with a concentration of 10^{18} cm^{-3} was up to 2 μm.

In connection with this, removal of the nucleation layer is an important technological step to achieve a uniform deposition along the whole depth of the porous layer. Figure 2c discloses the evolution of the nucleation layer and emergence of opened regions with pores disclosing self-ordering after isotropic wet etching for 10 s. The complete removal of the nucleation layer requires about 25 s (see Figure 2d). All further discussed electrochemical depositions were obtained in samples with opened pores obtained according to the procedure described in Section 2.2.

Adjusting the pulse width (t_{on}) and, especially, the delay between pulses (t_{off}), is essential for a controlled deposition of metals in semiconductor porous templates, as shown in Figure 3. The current pulses are optimized from the point of view of avoiding the electrolyte depletion inside pores and refreshing of the metal species during the delay between pulses.

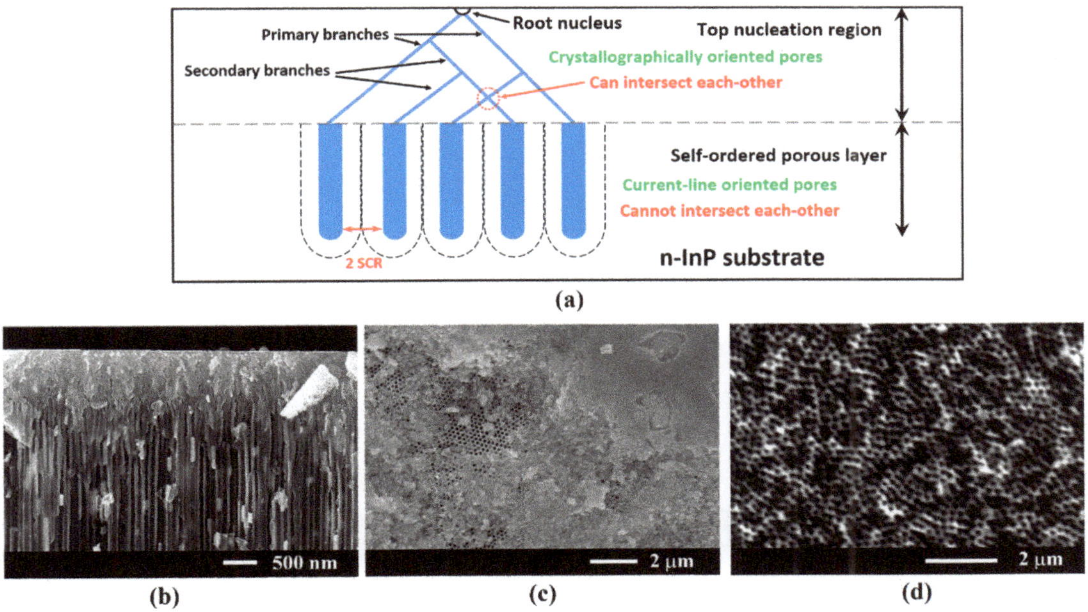

Figure 2. (**a**) Schematic representation of self-ordered porous layer formation; (**b**) cross-sectional SEM image of deposited segmented Pt nanotubes in porous InP template with a top nucleation layer ($U = -16$ V; $t_{on} = 100$ μs; $t_{off} = 1$ s); (**c**) top view SEM images after removal of the top nucleation layer by wet chemical etching in $HCl:H_3PO_4$ (1:1) during 10 s; (**d**) the same for etching during 25 s.

It was found that the pulse width should be set at a value enabling the deposition of only 70%–80% of the metal ions inside the pores during the applied pulse, in order to ensure an effective electrolyte refreshing and to avoid its complete depletion. Electrolyte depletion occurred if the pulse width was too long. In such a case, the metal ions managed to penetrate only a short distance inside the pores, from the template surface, which led to pore blocking and deposition on the template surface (Figure 3b).

Nevertheless, along with the optimized pulse width and the interval between pulses, the voltage pulse amplitude plays an important role. According to our previous study, by means of Raman scattering, the porous ZnSe skeleton showed a decrease in the free carrier concentration by a factor of two in comparison with the bulk ZnSe crystal, due to surface depletion effects which are enhanced in porous media with a large surface to-volume ratio [27,28]. The electrical conductivity of semiconductor nanostructures strongly depend upon their transversal dimensions, as was demonstrated by pulsed electrochemical deposition of gold on InP nanowalls, nanowires and nanobelts [29]. Taking into account these findings and the preferential deposition in more conductive regions, one can conclude that site-selective deposition of metallic nanostructures is achievable by adjusting the amplitude of current pulses. The distribution of current lines at the bottom of the porous/bulk interface when a low pulse voltage amplitude is applied, is illustrated by blue lines in Figure 3c (row II). In such a case, the deposition of metallic nanotubes or nanodots takes place at a specific depth of the porous layer according to the applied pulse voltage value. It is worth mentioning that long duration deposition in such conditions leads to the filling of pore from the bottom towards the pore entrance, therefore, resulting in the growth of metallic nanowires.

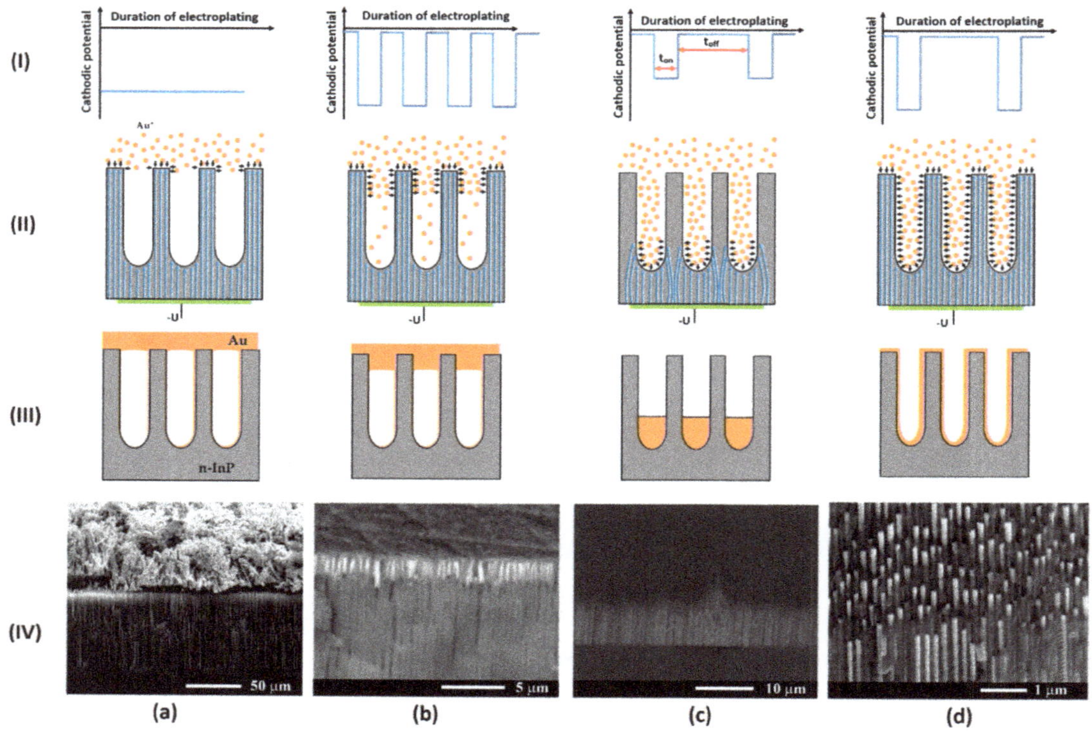

Figure 3. Site-selective pulsed electroplating of metal species in porous InP semiconductor nanotemplates at different deposition parameters: (**a**) continuously applied voltage leading to deposition on top of the template ($t_{off} = 0$); (**b**) deposition on and underneath surface at small delay applied between the pulses (t_{off}); (**c**) selective deposition of metal at the bottom of pores by applying a small voltage pulse amplitude; and (**d**) uniform deposition along the whole template thickness at optimized pulse width, delay between pulses and voltage amplitude resulting in the formation of Pt nanotubes. In row (**I**), a schematic representation of applied pulses is presented. In row (**II**), a schematic art-view of the metal species penetration inside pores, depending on the pulse parameters is presented; the blue lines illustrate the current lines. In row (**III**), a schematic representation of the deposited metal in porous InP template is presented, and row (**IV**) demonstrates the site-selective deposition by SEM images in cross-sectional view.

Another important parameter in pulsed electrochemical deposition designated for obtaining metal nanodots or nanotubes, is the total duration of electroplating (t_{dep}). The so-called "hopping electrodeposition" mechanism of gold nanodots on porous semiconductor structures was proposed and demonstrated in a previous work [30]. It was demonstrated that after nucleation, each dot increases in size up to a critical transverse dimension of about 20 nm, followed by nucleation of another gold nanodot in the vicinity. The process continues until the entire surface exposed to the electrolyte is covered by a monolayer of self-assembled gold nanodots forming walls of metal nanotubes (Figure 4a). In a longer duration of electroplating, the walls of the metal nanotubes become thicker, as can be seen in Figure 4b. The optimum parameters of the pulses (Figure 3d) assure the uniform deposition along the nanotube's depth.

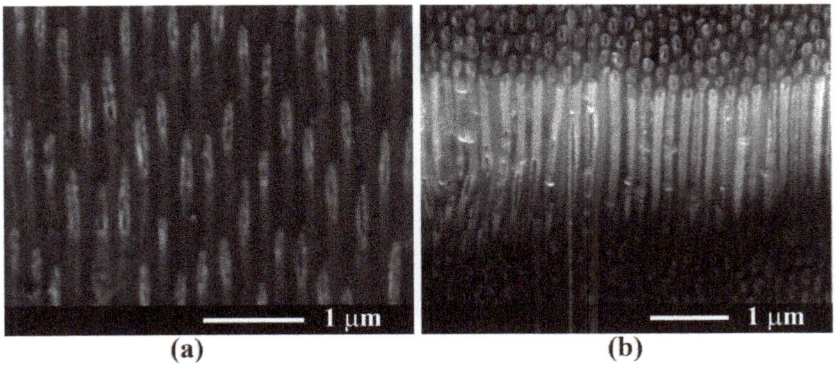

Figure 4. SEM images in cross-section, of Pt nanotubes in semiconductor InP envelope after electrochemical deposition with electroplating parameters of U = −16 V, t_{on} = 100 μs, t_{off} = 1 s, at different durations of electroplating: (**a**) 1 h; and (**b**) 2.5 h.

The site-selective deposition of Pt in a porous GaP template was applied for the fabrication of a variable capacitance device with Schottky contact (Figure 5). Uniform Pt nanotubes were deposited in template pores at a depth of 70 μm as the first step of the electrodeposition process with optimized pulse parameters (t_{off} = 1 s), according to Figure 3d, while the up contact of the device was formed at the second step of the electrodeposition process with pulse parameters leading to electrolyte depletion inside the pores (t_{off} = 10 ms), according to Figure 3b.

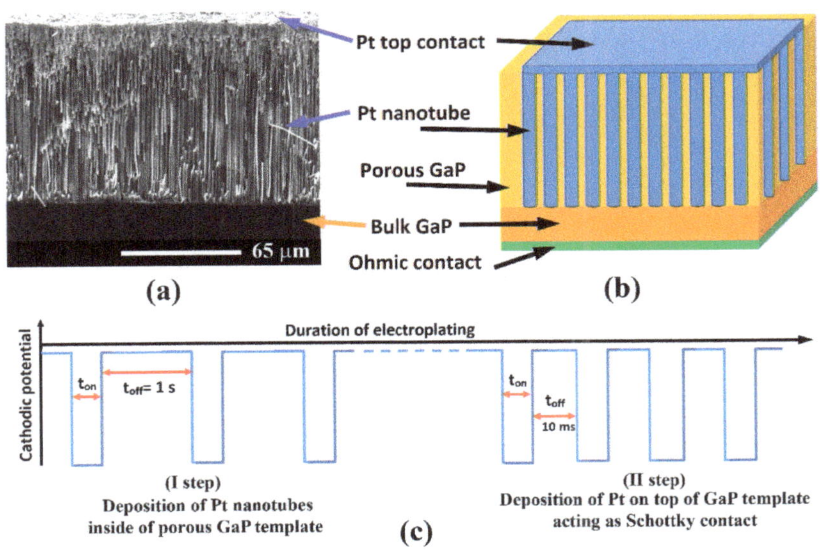

Figure 5. (**a**) SEM image in cross-section, of 100 μm deep porous GaP/Pt template; (**b**) the schematic presentation of the varicap device structure; (**c**) the sequence of pulses and pause between them during Pt electroplating in porous GaP template for Pt nanotubes and top contact deposition in the same technological process.

The electrochemical deposition of the up contact in the same technological process is an advantage compared with evaporation techniques, since it assures the interconnection of all Pt nanotubes leading to better performance of the varicap device. A record gradient of capacity per square micrometer of the template-based variable capacitance device was reported [31,32].

Our estimations suggest that increasing the pore depth in the porous GaP template up to 300 μm would improve the device performance up to four to five times. Preliminary results indicate that increasing the pause between pulses to 3 s, and decreasing the pulse width from 300 μs to 100 μs under permanent magnetic steering, is required to reach this goal.

As mentioned in the introduction section, controlled electrodeposition in porous oxide templates is challenging because of its high electrical resistivity. One can propose an alternative approach for the preparation of noble metal deposition inside oxide templates based on a two-step process: metals are deposited in porous semiconductor templates in the first step, as discussed above for InP templates; then, after noble metal deposition, the semiconductor skeleton is oxidized. This approach was demonstrated in the instance of porous GaAs templates in the form of arrays of nanowires, as shown in Figure 6. Arrays of GaAs nanowires were produced by anodization of GaAs(111)B wafers in an 1M HNO_3 electrolyte at 4 V for 20 min (Figure 6a). Au nanodots were deposited on the GaAs array by electroplating at pulse voltage value −16 V, t_{on} = 100 μs, t_{off} = 1 s, for 300 s (Figure 6b). Oxidation of GaAs wires after electroplating was performed by annealing at 800 °C for 1 h in air. Figure 6c,d shows the results of EDX analysis before and after oxidation, respectively.

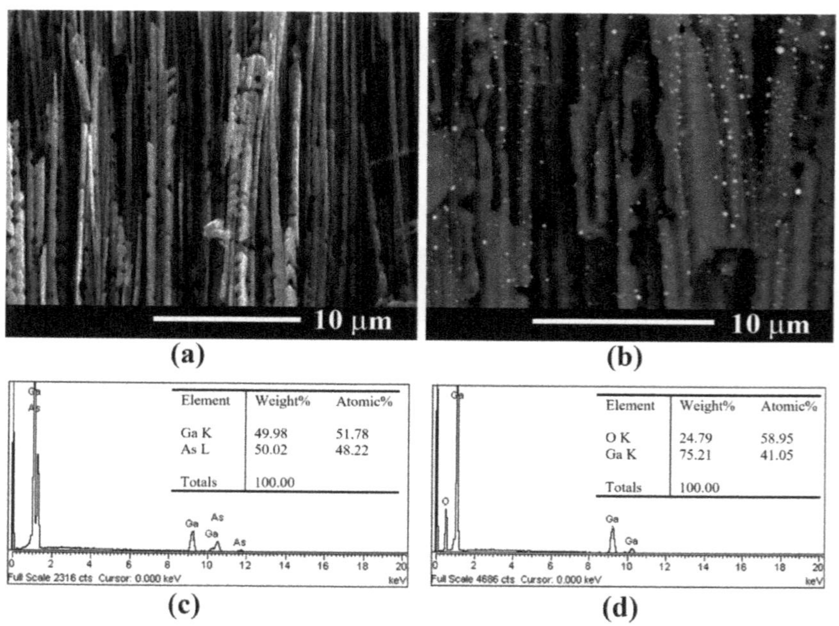

Figure 6. (**a**) GaAs nanowires after anodization of GaAs(111)B crystal in 1M HNO_3 at 4 V for 20 min; (**b**) Ga_2O_3/Au nanowires obtained by annealing at 800 °C for 1 h of GaAs nanowires with deposited gold nanodots; (**c**) the results of chemical composition analysis of GaAs nanowires after anodization; (**d**) the same after thermal annealing.

Figure 7 illustrates arrays of InP nanowires covered by Pt coatings. InP nanowires were produced by fast electrochemical etching [33] of InP substrates with an electron concentration of 1.3×10^{18} cm^{-3} in a 5% HCl electrolyte at an anodization voltage of 15 V for 3 s (Figure 7a). The approach of "fast electrochemical etching" was used, 2-μm long InP nanowires being obtained in just 3 s of anodization. Note that anodization at 5–6 V results in the formation of self-ordered porous InP structure, while ultrathin nanowalls, nanobelts or nanowires are produced at a relatively high applied potential, leading to the electrochemical dissolution of InP crystal around the remained nanostructures [33]. This approach, in combination with pulsed electrochemical deposition of gold nanodots, was used to assess the conductivity of the fabricated InP nanostructures depending on their thickness [29]. On the other hand, core–shell InP/Pt structures were fabricated by electroplating Pt on the prepared InP nanowires at a pulse voltage value −16 V, t_{on} = 300 μs, t_{off} = 0.5 s, for 1.5 h (Figure 7b). It should be noted that a uniform Pt deposition along the nanowires took place and interruptions occurred due to cross-sectioning of the sample.

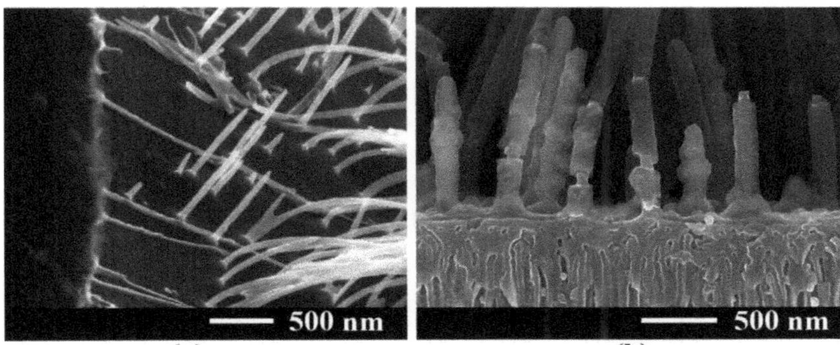

Figure 7. SEM cross-sectional image of: (**a**) InP nanowires obtained via fast electrochemical etching of InP substrates at 15 V for 3 s; (**b**) core–shell InP/Pt structure on porous InP.

4. Conclusions

The results of this study demonstrate possibilities for site-selective pulsed electrodeposition of metallic nanostructures in porous templates prepared by electrochemical etching of semiconductor wafers. Metal deposition occurred in the upper part of pores near the porous template surface under conditions of high amplitude, long current pulse width, and short interval between pulses, while the metal was deposited at the bottom of pores at conditions of low pulse amplitude and long intervals between pulses. Uniform deposition of the metal along the whole pore length was obtained with pulse parameters optimized from the point of view of effective electrolyte refreshing and avoiding its depletion in pores. This technology can be applied for the development of optoelectronic devices, particularly variable capacitance devices, as well as for plasmonic and photocatalytic applications. The technology can also be applied for the preparation of metal nanostructures on porous oxide templates, when it is combined with thermal treatment for the oxidation of the porous semiconductor skeleton.

Author Contributions: Conceptualization, E.V.M.; methodology, E.I.M., V.V.U., E.V.M.; validation, V.V.U. and E.V.M.; investigation, E.I.M. and E.V.M.; writing—original draft preparation, V.V.U. and E.V.M.; writing—review and editing, V.V.U., E.V.M. and I.M.T.; visualization, E.I.M. and E.V.M.; supervision, I.M.T.; project administration, E.V.M. and I.M.T.; funding acquisition, I.M.T. All authors have read and agreed to the published version of the manuscript.

Funding: This research was funded by National Agency for Research and Development of Moldova under grants #20.80009.5007.20 and #21.00208.5007.15/PD "Micro- and nano-engineering of semiconductor compounds based on electrochemical technologies for electronic and photonic applications". The APC was funded by European Commission under the H2020 grant #810652 'NanoMedTwin'.

Institutional Review Board Statement: Not applicable.

Informed Consent Statement: Not applicable.

Data Availability Statement: The data presented in this study are available on request from the corresponding author.

Acknowledgments: E.V.M. acknowledges support from the Alexander von Humboldt Foundation (Bonn, Germany).

Conflicts of Interest: The authors declare no conflict of interest. The founding sponsors had no role in the design of the study; in the collection, analyses, or interpretation of data; in the writing of the manuscript, and in the decision to publish the results.

References

1. Anik, M.I.; Hossain, M.K.; Hossain, I.; Mahfuz, A.M.U.B.; Rahman, M.T.; Ahmed, I. Recent Progress of Magnetic Nanoparticles in Biomedical Applications: A Review. *Nano Sel.* **2021**, *2*, 1146–1186. [CrossRef]
2. Li, K.; Xu, J.; Li, P.; Fan, Y. A Review of Magnetic Ordered Materials in Biomedical Field: Constructions, Applications and Prospects. *Compos. Part B Eng.* **2022**, *228*, 109401. [CrossRef]
3. Cardoso, V.F.; Francesko, A.; Ribeiro, C.; Bañobre-López, M.; Martins, P.; Lanceros-Mendez, S. Advances in Magnetic Nanoparticles for Biomedical Applications. *Adv. Healthc. Mater.* **2018**, *7*, 1700845. [CrossRef]
4. Mohammed, H.; Moreno, J.A.; Kosel, J. *Advanced Fabrication and Characterization of Magnetic Nanowires*; IntechOpen: London, UK, 2017; ISBN 978-1-78923-679-8.
5. Shah, K.W.; Xiong, T. Multifunctional Metallic Nanowires in Advanced Building Applications. *Materials* **2019**, *12*, 1731. [CrossRef] [PubMed]
6. Proenca, M.P.; Sousa, C.T.; Ventura, J.; Araújo, J.P. 6—Cylindrical Magnetic Nanotubes: Synthesis, Magnetism and Applications. In *Magnetic Nano- and Microwires*, 2nd ed.; Vázquez, M., Ed.; Woodhead Publishing Series in Electronic and Optical Materials; Woodhead Publishing: Sawston, UK, 2020; pp. 135–184. ISBN 978-0-08-102832-2.
7. Staňo, M.; Fruchart, O. Chapter 3—Magnetic Nanowires and Nanotubes. In *Handbook of Magnetic Materials*; Brück, E., Ed.; Elsevier: Amsterdam, The Netherlands, 2018; Volume 27, pp. 155–267.
8. Sharma, M.; Kuanr, B.K. *Electrodeposition of Ferromagnetic Nanostructures*; IntechOpen: London, UK, 2015; ISBN 978-953-51-2213-5.
9. Gerein, N.J.; Haber, J.A. Effect of AC Electrodeposition Conditions on the Growth of High Aspect Ratio Copper Nanowires in Porous Aluminum Oxide Templates. *J. Phys. Chem. B* **2005**, *109*, 17372–17385. [CrossRef] [PubMed]
10. Yue, S.; Zhang, Y.; Du, J. Preparation of Anodic Aluminum Oxide Tubular Membranes with Various Geometries. *Mater. Chem. Phys.* **2011**, *128*, 187–190. [CrossRef]
11. Enculescu, I. Nanowires and Nanotubes Prepared Using Ion Track Membrane Templates. *Dig. J Nanomater. Biostructures* **2006**, *1*, 15–20.
12. Enculescu, I.; Siwy, Z.; Dobrev, D.; Trautmann, C.; Toimil Molares, M.E.; Neumann, R.; Hjort, K.; Westerberg, L.; Spohr, R. Copper Nanowires Electrodeposited in Etched Single-Ion Track Templates. *Appl. Phys. A* **2003**, *77*, 751–755. [CrossRef]
13. Martin, C.R. Nanomaterials: A Membrane-Based Synthetic Approach. *Science* **1994**, *266*, 1961–1966. [CrossRef]
14. Sulka, G.; Zaraska, L.; Stępniowski, W. Anodic Porous Alumina as a Template for Nanofabrication. In *Encyclopedia of Nanoscience and Nanotechnology*; American Scientific Publishers: Valencia, CA, USA, 2011; Volume 11, pp. 261–349.
15. Tiginyanu, I.; Ursaki, V.; Monaico, E. Template Assisted Formation of Metal Nanotubes. In *Nanostructures and Thin Films for Multifunctional Applications: Technology, Properties and Devices*; Tiginyanu, I., Topala, P., Ursaki, V., Eds.; NanoScience and Technology; Springer International Publishing: Cham, Switzerland, 2016; pp. 473–506. ISBN 978-3-319-30198-3.
16. Langa, S.; Carstensen, J.; Tiginyanu, I.M.; Christophersen, M.; Föll, H. Formation of Tetrahedron-Like Pores during Anodic Etching of (100) Oriented n-GaAs. *Electrochem. Solid-State Lett.* **2001**, *5*, C14. [CrossRef]
17. Monaico, E.; Tiginyanu, I.; Ursaki, V. Porous Semiconductor Compounds. *Semicond. Sci. Technol.* **2020**, *35*, 103001. [CrossRef]
18. Tiginyanu, I.M.; Ursaki, V.V.; Monaico, E.; Enachi, M.; Sergentu, V.V.; Colibaba, G.; Nedeoglo, D.D.; Cojocaru, A.; Föll, H. Quasi-Ordered Networks of Metal Nanotubes Embedded in Semiconductor Matrices for Photonic Applications. *J. Nanoelectron. Optoelectron.* **2011**, *6*, 463–472. [CrossRef]
19. Belov, A.N.; Gavrilov, S.A.; Shevyakov, V.I.; Redichev, E.N. Pulsed Electrodeposition of Metals into Porous Anodic Alumina. *Appl. Phys. A* **2011**, *102*, 219–223. [CrossRef]
20. Abbondanza, G.; Larsson, A.; Linpé, W.; Hetherington, C.; Carlá, F.; Lundgren, E.; Harlow, G.S. Templated Electrodeposition as a Scalable and Surfactant-Free Approach to the Synthesis of Au Nanoparticles with Tunable Aspect Ratios. *Nanoscale Adv.* **2022**, *4*, 2452–2467. [CrossRef] [PubMed]
21. Tiginyanu, I.; Monaico, E.; Monaico, E. Ordered Arrays of Metal Nanotubes in Semiconductor Envelope. *Electrochem. Commun.* **2008**, *10*, 731–734. [CrossRef]
22. Dela Pena, E.M.; Roy, S. Electrodeposited Copper Using Direct and Pulse Currents from Electrolytes Containing Low Concentration of Additives. *Surf. Coat. Technol.* **2018**, *339*, 101–110. [CrossRef]
23. Boukhouiete, A.; Creus, J. Nickel Deposits Obtained by Continuous and Pulsed Electrodeposition Processes. *J. Mater. Environ. Sci.* **2015**, *6*, 1840–1844.
24. Renner, F.U.; Ankah, G.N.; Bashir, A.; Ma, D.; Biedermann, P.U.; Shrestha, B.R.; Nellessen, M.; Khorashadizadeh, A.; Losada-Pérez, P.; Duarte, M.J.; et al. Star-Shaped Crystallographic Cracking of Localized Nanoporous Defects. *Adv. Mater.* **2015**, *27*, 4877–4882. [CrossRef]

25. Plesco, I.; Ciobanu, V.; Braniste, T.; Ursaki, V.; Rasch, F.; Sarua, A.; Raevschi, S.; Adelung, R.; Dutta, J.; Tiginyanu, I. Highly Porous and Ultra-Lightweight Aero-Ga2O3: Enhancement of Photocatalytic Activity by Noble Metals. *Materials* **2021**, *14*, 1985. [CrossRef]
26. Enachi, M.; Guix, M.; Postolache, V.; Ciobanu, V.; Fomin, V.M.; Schmidt, O.G.; Tiginyanu, I. Light-Induced Motion of Microengines Based on Microarrays of TiO2 Nanotubes. *Small* **2016**, *12*, 5497–5505. [CrossRef]
27. Monaico, E.; Tiginyanu, I.M.; Ursaki, V.V.; Sarua, A.; Kuball, M.; Nedeoglo, D.D.; Sirkeli, V.P. Photoluminescence and Vibrational Properties of Nanostructured ZnSe Templates. *Semicond. Sci. Technol.* **2007**, *22*, 1115–1121. [CrossRef]
28. Irmer, G.; Monaico, E.; Tiginyanu, I.M.; Gärtner, G.; Ursaki, V.V.; Kolibaba, G.V.; Nedeoglo, D.D. Fröhlich Vibrational Modes in Porous ZnSe Studied by Raman Scattering and Fourier Transform Infrared Reflectance. *J. Phys. Appl. Phys.* **2009**, *42*, 045405. [CrossRef]
29. Monaico, E.V.; Tiginyanu, I.M.; Ursaki, V.V.; Nielsch, K.; Balan, D.; Prodana, M.; Enachescu, M. Gold Electroplating as a Tool for Assessing the Conductivity of InP Nanostructures Fabricated by Anodic Etching of Crystalline Substrates. *J. Electrochem. Soc.* **2017**, *164*, D179. [CrossRef]
30. Tiginyanu, I.; Monaico, E.; Nielsch, K. Self-Assembled Monolayer of Au Nanodots Deposited on Porous Semiconductor Structures. *ECS Electrochem. Lett.* **2015**, *4*, D8. [CrossRef]
31. Tiginyanu, I.; Monaico, E.; Ursaki, V. Two-Dimensional Metallo-Semiconductor Networks for Electronic and Photonic Applications. *ECS Trans.* **2012**, *41*, 67. [CrossRef]
32. Tiginyanu, I.; Monaico, E.; Sergentu, V.; Tiron, A.; Ursaki, V. Metallized Porous GaP Templates for Electronic and Photonic Applications. *ECS J. Solid State Sci. Technol.* **2015**, *4*, P57. [CrossRef]
33. Monaico, E.; Tiginyanu, I.; Volciuc, O.; Mehrtens, T.; Rosenauer, A.; Gutowski, J.; Nielsch, K. Formation of InP Nanomembranes and Nanowires under Fast Anodic Etching of Bulk Substrates. *Electrochem. Commun.* **2014**, *47*, 29–32. [CrossRef]

coatings

Article

Transforming Chimney Soot via Stochastic Polymerization for Active Electrode Coating

Miroslav Petrov [1,*], Lyubomir Slavov [1], Toma Stankulov [2], Boryana Karamanova [2], Teodor Milenov [1], Dimitar Dimov [1] and Ivalina Avramova [3]

[1] Institute of Electronics, Bulgarian Academy of Sciences, 72 Tsarigradsko Shaussee, 1784 Sofia, Bulgaria; l_slavov@ie.bas.bg (L.S.); tmilenov@ie.bas.bg (T.M.); dadimov@ie.bas.bg (D.D.)
[2] Institute of Electrochemistry and Energy Systems, Bulgarian Academy of Sciences, Academician Georgi Bonchev Street, 1113 Sofia, Bulgaria; tstankulov@iees.bas.bg (T.S.); boriana.karamanova@iees.bas.bg (B.K.)
[3] Institute of General and Inorganic Chemistry, Bulgarian Academy of Sciences, Academician Georgi Bonchev Street, 1113 Sofia, Bulgaria; iva@svr.igic.bas.bg
* Correspondence: petrov80@abv.bg or petrov_80@ie.bas.bg

Abstract: A polymerization procedure is presented to increase the molecular weight of hydrocarbons in household chimney soot without thermal treatment at high temperatures. Pristine soot was subject to chlorination, with half of it treated with magnesium (Mg-plates) to create random-type Grignard reagents (R-Mg-Cl) in diethyl ether media. Mixing the Grignard reagent and the rest of the halogenated soot material created new C-C bonds, thus increasing the molecular weight of the final product. The obtained stochastically polymerized soot (SPS) was investigated using Raman spectroscopy, FTIR spectroscopy and XPS and was subjected to electrochemical testing as an assembled supercapacitor with a KOH electrolyte. Results show significant carbon structure differences due to the chemical procedures and newly created functional groups in the soot. Such functional groups could increase the capacity of supercapacitors, creating pseudo-capacitance by participating in redox reactions. The results also unveiled removing any random contaminations in the pristine soot and obtaining a more uniform final product containing hydrocarbons with longer chains, thus increasing the molecular weight.

Keywords: soot; circle economy; waste treatment; super capacitors; organic synthesis

Citation: Petrov, M.; Slavov, L.; Stankulov, T.; Karamanova, B.; Milenov, T.; Dimov, D.; Avramova, I. Transforming Chimney Soot via Stochastic Polymerization for Active Electrode Coating. *Coatings* **2023**, *13*, 1354. https://doi.org/10.3390/coatings13081354

Academic Editor: Peter Rodič

Received: 25 June 2023
Revised: 25 July 2023
Accepted: 28 July 2023
Published: 2 August 2023

1. Introduction

The exceptional importance of the combustion of carbon-containing compounds as a source of thermal energy throughout human history and its related progress cannot be denied [1]. However, this process, essentially a reaction leading to numerous uncontrolled chemical changes in the starting hydrocarbons, is also associated with the production of harmful emissions into the atmosphere, including soot aerosols [1–3].

The incomplete combustion of hydrocarbons associated with both anthropogenic (various combustion systems for energy and heat production, transport, etc.) and natural sources (forest fires) is responsible for the formation and emission of more than 8 Tg of soot annually into the atmosphere [2–7]. With a more than 80% contribution, elemental carbon is the main constituent of soot particles regardless of the source [8–10]. The processes of the formation of soot, which passes roughly through four stages, i.e., from precursor formation to aggregate formation, have been the subject of many notable studies, summarized in a recent review article by Martin J. et al. [9]. The final product in the form of carbonaceous particulate matter, both as aerosols and as solid deposits (in diesel and gasoline particle filters, onto chimneys' internal walls, etc.), leads to serious health and environmental risks for the population [7,9,11]. The industry has made enormous efforts to reduce the soot particles released into the atmosphere. But industrial advances can also lead to additional increases in the amount of solid deposits of soot and the generation of unnecessary waste products. At the same time, they could lead to obtaining even more soot as solid deposits, which indeed results in a huge amount of unnecessary cheap waste product. For instance, in the context of engine-generated soot emissions, besides advances in particle filters [12] and modifications in the combustion process like the use of EGR (exhaust gas recirculation), attempts were made to lower the temperature of the exhaust gases [13,14]. However, the EGR leads to higher quantities of soot in the crankcase oil [13], and lowering the temperatures leads to an increased rate of soot deposition [14]. To these amounts of soot as waste material must be added those obtained during the incomplete combustion of hydrocarbons in households. The study of the physicochemical properties of soot and the produced materials from them is of great importance for establishing their true value for practical applications, since a key point in the conceptual model of a circular economy is the successful application of waste materials as useful ones [3,5,10,15].

Although soot (as carbon black) has been widely used as a component in inks since ancient times [16], a recent review focused on the currently known applications of recycled soot [17] suggests the enormous, still-untapped potential of this abundant waste material. To the soot applications discussed by Uttaravalli et al. [17], some other studies on soot as a material for anti-bio-adhesion behavior coatings [18], as super hydrophobic coatings [19], and also some experimental attempts for soot to be used as supercapacitor electrodes [10,20,21] should be added. Functionalized candle soot is used to improve Raman sensing properties for active SERS (surface-enhanced Raman scattering) substrates [22].

On the other hand, to achieve specific physical properties and specialized applications for advanced structures and architectures, one could rely on the constant progress in polymer science [23–25]. Extensive research in the field of carbocationic and free radical polymerization (FRP) led to the synthesis of new well-defined (co)polymers, with FRP already being probably the most important commercial process, which allows one to obtain high molar mass polymers [23–27]. The discovery of Grignard agents could also be considered one important stage of development in polymer science. The ease of preparation of these organo-magnesium reagents, discovered at the beginning of the last century by Victor Grignard, ensures their large-scale success and ensures their wide application in organic and organometallic synthesis [28]. The reaction to obtain Grignard reagents (R-Mg-X) in various solvents is generally described as R-X + Mg, where R is a carbon-containing radical and X is a halogen element [28,29]. These reagents react with numerous organic functional groups containing polar multiple bonds (ketones, nitriles, sulfones, etc.), highly strained carbon rings, acidic hydrogens, etc. [29].

In the present study, a polymerization procedure was applied to alter the properties of what would otherwise be waste products (household chimney soot) in a way in which they could be further used as material for supercapacitance applications. Since the supercapacitors are a unique class of energy-storage devices that are mainly produced from carbon-based materials [30–32], attempts at this scientific direction should be regarded as needed steps toward the realization of a true circular economy [15]. Probably equally common and harmful as all other types of soot, also easily accessible and costless, but at the same time structurally different because of the specific burning conditions, household chimney soot lags far behind in terms of scientific research compared with all other types of carbonaceous materials.

Stochastic polymerization appears when complicate materials are involved in polymerization processes and different exchanging segments and molecules are interacting [33]. Here, stochastic free radical synthesis, involving a combination of newly created Grignard reagent and chlorinated soot, leads to the creation of new higher molecular compounds in pristine soot.

The novel chemical routine is described in detail. Raman spectroscopy, Fourier-transform infrared spectroscopy (FTIR) spectroscopy and X-ray photoelectron spectroscopy (XPS) analysis were applied as techniques to characterize the obtained stochastically polymerized soot (SPS). The material was applied as an active electrode coating and was subjected to electrochemical testing with KOH as an electrolyte.

2. Materials and Methods

2.1. Experimental Section

Swept household chimney soot was used in the experimental procedure, shown as block scheme no. Figure 1 and described in detail below. Procedure must be performed carefully because of the starting exothermic reaction and the occurrence of immense amount of fine, thick foam. To avoid evaporation of chlorinated hydrocarbons, the mixture thus obtained was dried in a desiccator containing crystal $CaCl_2$ and KOH for a month at RT.

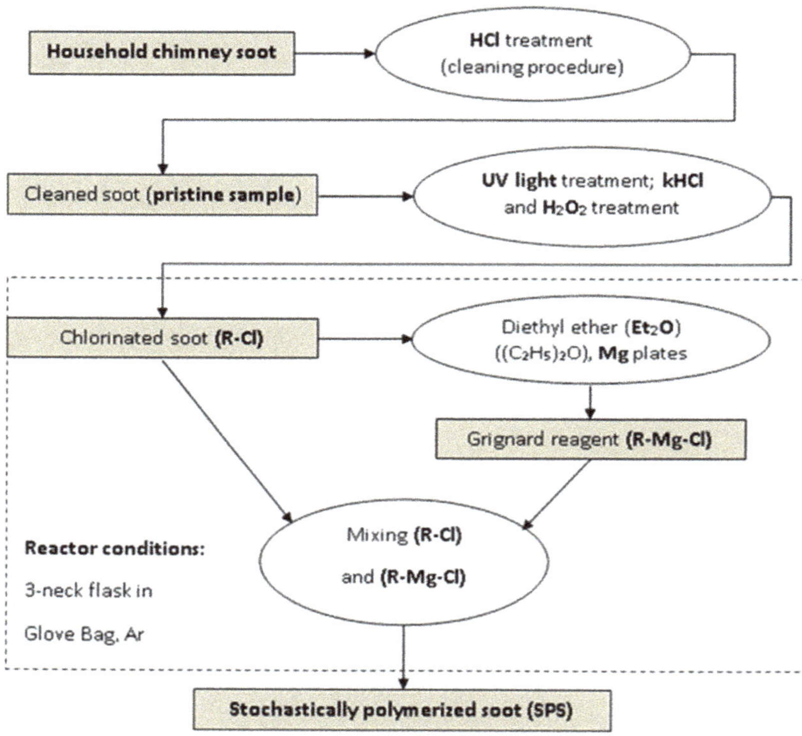

Figure 1. Block-scheme of the experimental procedures involved in altering household chimney soot into stochastically polymerized soot (SPS).

After such a prolonged period, it was expected that all the moisture, CO_2, and surplus HCl reacted with the $CaCl_2$ and KOH crystallites.

The experimental procedures on the chlorinated soot (ChS) material continued in a 3-neck 200 mL flask, used as a reactor to obtain Grignard reagent (R-Mg-Cl). One of the three necks of the reactor was for inserting a glass rod used as a stirrer, the second one was used as a breather (filled with crystal $CaCl_2$ between cotton tampons) and the central neck was used for filing the flask. The flask was rinsed with diethyl ether ($(C_2H_5)_2O$) or Et_2O) prior to experiment and placed in an ice water bath at 0 °C. Every glass piece of the equipment was heated for 2 h at 110 °C prior to experiment. Two 15 mL vessels filled with powdered ChS in each, 100 mL Et_2O and 1 g Mg (fine plates) were placed in a Glove Bag filled with inert gas (Ar). The ChS from one of the vessels was placed in the flask, followed by the 100 mL Et_2O and the Mg plates. The filling neck of the flask was capped and the

Glove Bag was removed. This was followed by vigorous stirring of the solution, aiming at the maximum dissolution of the magnesium plates. A brief heating of the mixture to ~30 °C resulted in slight boiling (due to the low boiling point of the ether in the solution), then the mixture was quenched in the ice bath. After 12 h, needed for optimal reaction result, the Glove Bag again covered the installation and purged with Ar. A brief heating (to ~30 °C) was applied to increase the partial pressure in the volume, preventing of inlet of external gases. A separation funnel with a long glass tube at the bottom end and a syringe at the top end was used to evict the obtained Grignard reagent upwards via the creation of a vacuum. The unreacted Mg plates and sediment that remained at the bottom were removed through the breather neck of the flask, and the flask was rinsed again with ether. The second vessel with powdered chlorinated soot was then placed in the flask, and 20 mL of ether and the Grignard reagent were also added. After 24 h, the liquid phase was removed from the obtained mixture using a pipette, and the precipitated material was collected and dried at 30 °C for 1 h, resulting in the final experimental product: stochastically polymerized soot (SPS).

2.2. Methods of Investigation

The Raman spectra were taken using HORIBA Jobin Yvon Labram HR 800 spectrometer equipped with a Peltier-cooled CCD detector with He-Ne laser excitation of 633 nm wavelength and laser power at 0.5 mW. The spectral resolution was 1 cm^{-1} or better. The materials were also investigated using Fourier-transform infrared spectroscopy (FTIR) spectroscopy on Tenzor 27-Bruker apparatus in the range of 4000–550 cm^{-1}. The X-ray photoelectron spectroscopy (XPS) studies were performed in a VG Escalab MKII electron spectrometer using monochromatic AlKα radiation with an energy of 1486.6 eV under base pressure 10^{-8} Pa and a total instrumental resolution 1 eV. The binding energies (BEs) were determined utilizing the C 1s line as a reference with an energy of 285.0 eV. The accuracy of the measured BE was 0.2 eV. The photoelectron lines of constituent elements on the surface were recorded and corrected by subtracting a Shirley-type background and quantified using the peak area and Scofield's photoionization cross-sections. The deconvolution of spectra was performed with XPSPEAK41 software (version 4.1). The conditions used for electrochemical testing are presented in Section 3.4.

3. Results and Discussion

3.1. Raman Spectroscopy Investigation

Raman spectroscopy investigation was performed on the pristine soot sample and the SPS sample to unveil how the chemical treatment influences the pristine carbon structure in the soot (Figure 2).

Since carbon is the main constituent in the material, it is not surprising that Raman spectroscopy is a widely used method for the characterization of any type of soot [3,5,6,8,12,13,34–42].

In the first-order spectral region (1000–2000 cm^{-1}), both spectra in Figure 2 exhibited the two typical-for-all-carbon-species broad bands, centered at 1364 cm^{-1} and 1583 cm^{-1}, corresponding to the D- and G-band, respectively. A detailed description of the significance of these two peaks for the state of the carbon structure can be found in the seminal work of Ferrari and Robertson [43].

While the D peak was related to defects in the ordered graphite structure (crystalline defects, vacancies, etc.), the G peak was regarded as an expression of that same ordered graphitic structure [40]. Regarding soot samples, the G peak signaled some local order associated with the presence of sp^2 aromatic carbon rings in the predominantly disordered structure [12,36]. In the spectral region of 2100–2800 cm^{-1}, the pristine soot spectra exhibited a wide "bump" which comprised all overtone modes associated with the first-order carbon peaks (D and G).

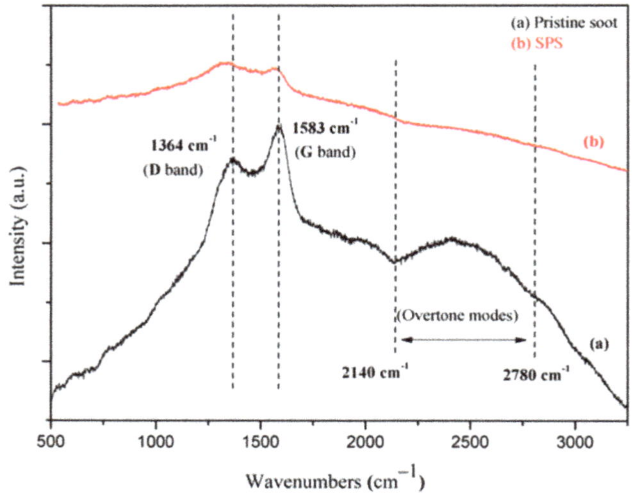

Figure 2. Raman spectra of (a) pristine soot and (b) stochastically polymerized soot (SPS).

But, these second-order peaks are rarely reported in the literature when regarding soot samples [44], so a contribution from some organic carbon (OC) or hydrocarbons could be regarded in that region. The spectrum of the SPS sample flattened in that region. A second major difference between the two spectra is the change in places in terms of the intensity of the G- and D-bands: in the SPS sample, the D-band increased in intensity, while the G-band decreased compared with the pristine soot spectrum. The drastic decrease in background fluorescence in the final product spectrum, compared with the pristine soot spectrum (Figure 2), should also be noted. The steep slope of the experimental curve until reaching the first peak at 1356 cm^{-1} (line "a" in Figure 2) is a typical indicator for the presence of an interfering fluorescence signal (FS), alongside the very shallow plateau between the D- and the G- band. The FS in a soot spectrum is due to organic compounds like PAHs (polyaromatic hydrocarbons) and unsaturated hydrocarbons condensed onto the particles, which are normally produced during combustion [5,37,40,44]. Thus, when comparing the two spectra, it could be said that the decrease in the intensity of the fluorescent signal, along with the complete alignment of the SPS spectrum in the region 2100–2800 cm^{-1}, indicates the removal of any organic compounds and the achievement of a more uniform final product (SPS).

In order to further analyze the changes that took place during the pristine soots' chemical treatment, a four-band fit was applied and is presented in Figure 3. Sadezky et al. were the first to establish a Raman deconvolution routine for carbonaceous materials, including soot, with a five-band fit involving four Lorentz-shaped curves (labeled as G, D1, D2 and D4) and one Gaussian-shaped curve (D3) [8]. Since then, the D2 curve has rarely been used when fitting soot samples [12,38]. The additional curves used for the spectrum analysis here, besides the G and D1 band, which match the D-band position and meaning, were the less-intense D3 and D4 bands. The D3 band, lying close to 1500 cm^{-1}, is associated with the presence of amorphous carbon [8,12,13]. This carbon fraction in the soot comes from organic molecules, functional groups, fragments, etc. [8,12,13]. The D4 band, lying close to 1180 cm^{-1}, is associated (when in soot) with ionic impurities and/or polyenes [8,13,42]. The decrease in intensity of the G-band in SPS compared with the pristine soot means raised disorder in the graphitic structure, which could be connected with the functionalization procedures conducted. The sp^2 carbon atoms from the aromatic rings available in the pristine sample, most probably by adding chlorine atoms, changed their hybridization into sp^3 and, thus, the G-band shrunk in intensity. Therefore, it can be concluded that the newly created functional groups in the soot do exist in the final product (SPS), which is of big importance for supercapacitance applications. In addition, there is also a significant increase in the contribution of both the amorphous carbon phase (D3 band) and that of ionic impurities (D4 band) in the SPS spectrum compared with that of the pristine soot sample (Figure 3). To obtain more information about the origin of these two additional phases' increases, a further FTIR investigation was conducted. Nevertheless, the soot polymerization procedures involved have a major impact on altering the arbon phase composition of the original waste material.

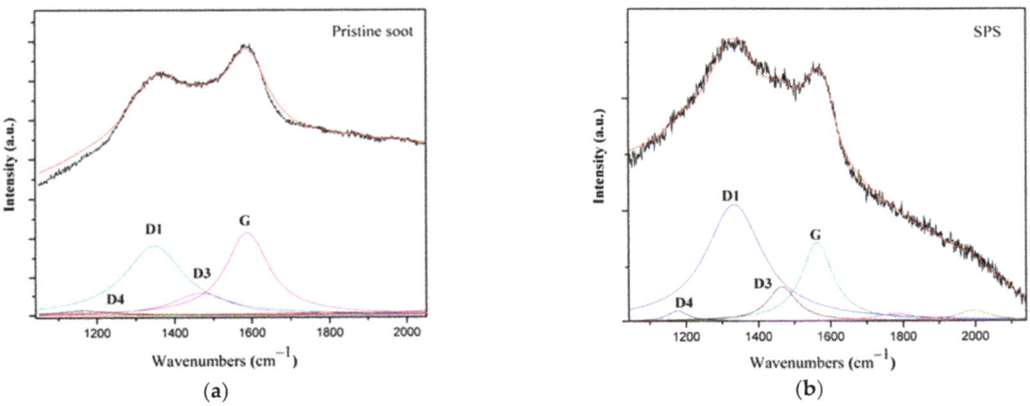

Figure 3. Four-band fit for the Raman spectra of (**a**) pristine soot and (**b**) stochastically polymerized soot (SPS).

3.2. FTIR Spectroscopy Investigation

FTIR spectra recorded on the pristine soot and the stochastically polymerized soot (SPS) materials are shown in Figure 4.

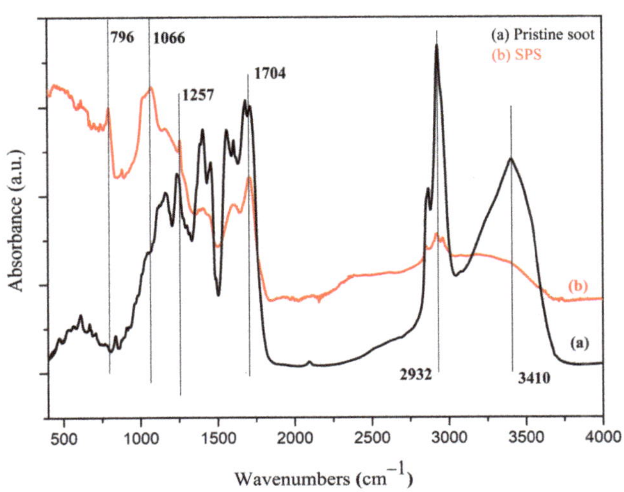

Figure 4. FTIR spectra of (a) pristine soot and (b) stochastically polymerized soot (SPS).

A significant change in the spectrum of the pristine soot's chemical composition is evident in Figure 4. Taking into consideration that any given absorption band intensity assigned to a functional group increases/decreases proportionately with the number of times that a functional group occurs within the molecule [45], the presented data confirm the results from the Raman spectroscopy.

Starting with the higher wavenumbers, it can be seen that the intensity of the peak around 3500 cm^{-1} (stretching vibrations, OH-groups) was significantly lower in the SPS sample's spectrum, compared with the pristine one (Figure 4). This means that OH groups were highly affected due to the chlorination process and the Grignard synthesis. In the spectral range 2960–2850 cm^{-1}, where the stretching vibrations of C-H alkane bonds are situated [46], a significant lowering of the intensity in the SPS spectrum was observed. This was mainly due to the chlorination process, where the chlorine atoms displaced the hydrogen atoms from the C-H bonds. The overall intensity decrease in the peaks in the SPS spectrum was, in general, until 1300 cm^{-1}, where the OH bending vibrations of phenol are situated (Figure 4). The peak at 1704 cm^{-1}, assigned to C=O-conjugated aldehyde or acid [46], decreased in intensity in the SPS spectrum, but was still significant in appearance. Under UV light, PAH and aromatic species with carbonyl and carboxyl groups generated excited species, which then underwent further reaction [7]. At the lower wavenumbers, the absorbance in the SPS spectrum, compared with the pristine soot spectrum, increased where the C-O, C-C and C=C bending vibrations of alkyl aryl ether, esters and alkenes appeared [45]. This means that new bonds were created, thus increasing the molecular weight of the soot in the SPS sample. A strong intensity peak emerged in the SPS spectrum at 1066 cm^{-1} related to C-O stretching vibrations [47] (Figure 4). The reason for its appearance may be due to the addition of H_2O_2 in the chlorination process. The hydrogen peroxide oxidized H_2, but H_2O_2 also reacted with carbon atoms and also could form additional C-O and C=O bonds. The existing C-O and C=O were probably oxidized to CO_2, which was the main phase of the emerging gases during chlorination.

Another peak also appeared at 796 cm^{-1} in the SPS spectrum, which, according to the Sigma Aldrich database table [46], is characteristic of C-H or C=C bending vibrations. In this case, the peak may be due to both types of characteristic bond vibrations. The overall increase in intensity below 1200 cm^{-1} in the SPS spectrum, compared with the pristine soot spectrum, most probably means that the final product has been subject to a reduction process. If a Grignard synthesis is performed with a significant amount of H_2O traces present, the water molecules destroy the Grignard reagent, thus giving simple C-H bonds instead of C-C bonds. However, the observed C-H and C-C spectral features in the FTIR spectra of the SPS sample had higher intensities, which is evidence that the polymerization was performed correctly, and hydrocarbons with longer chains were created.

3.3. XPS Analysis

The chemical state of the soot, PS and SPS samples were studied using X-ray Photoelectron Spectroscopy (XPS), presented in Figure 5. In the survey scan, clearly distinguished difference between the three spectra was that Cl was not present in the soot and PS samples. In addition, Si and Mg appeared in the SPS, which was due to the traces of Mg that remained after the synthesis, and the only source of Si was the used laboratory glass. There was N_2, of which occurrence is normal in the burning processes, but it disappeared in the final SPS sample.

The C 1s and Cl 2p deconvoluted photoelectron spectra of the pristine soot (PS) are shown in Figure 6.

The C 1s and Cl 2p peaks were deconvoluted to obtain information on the sp^2 and sp^3 carbon bonding fractions and oxygenated functional, as well as to evaluate the possible carbon-to-chlorine bonding. The C 1s peak for PS can be fitted well by three peaks: sp^2-hybridized carbon at 284.6 eV, sp^3-hybridized carbon at 285.2 eV and the C-OH group at 286.6 eV.

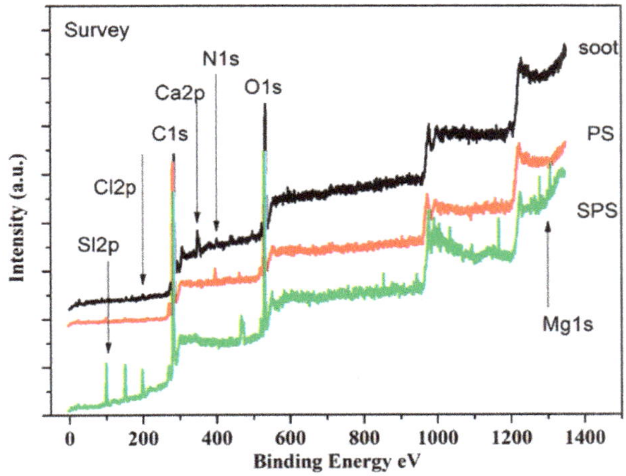

Figure 5. XPS spectra of soot, pristine soot (PS) and SPS.

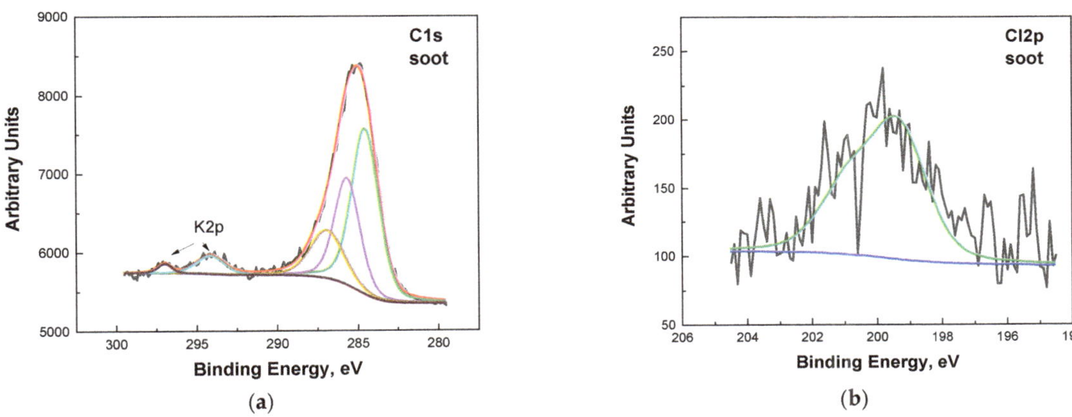

Figure 6. XPS Deconvolution spectrum taken from PS sample (**a**) C 1s line (**b**), Cl 2p line.

The C 1s for the SPS presented in Figure 7 was fitted by four components, respectively: sp^2-hybridized carbon at 284.6 eV, sp^3-hybridized carbon at 285.2 eV, the C-OH group at 286.6 eV and the C=O group at 288.8 eV. It is worth noting that K and Cl were present in small amounts in the PS sample; the K 2p and Cl 2p binding energies suggest that they were linked as KCl. The chemical routine caused some changes in the chemical composition and bonding in the SPS. The peak at 286.6 eV can be associated with the existence of C-Cl bonds [48–50], left from the ChS sample as well as C-OH groups, which were already present on the PS surface in a significant amount. The concentration of oxygen on the surface of the SPS was lowered by the same amount as the increased concentration of chlorine atoms.

The Cl 2p XPS spectra (a spin–orbit split into 2p 3/2 (200.7 eV) and 2p 1/2 (202.3 eV) peaks) for the SPS, shown in Figure 7b, is clear proof of the presence of chlorine atoms on the surface of the chlorinated sample. Additional conformation for that is the shift of Cl 2p binding energy from 199.3 eV to 200.7 eV, which is typical for organic compounds.

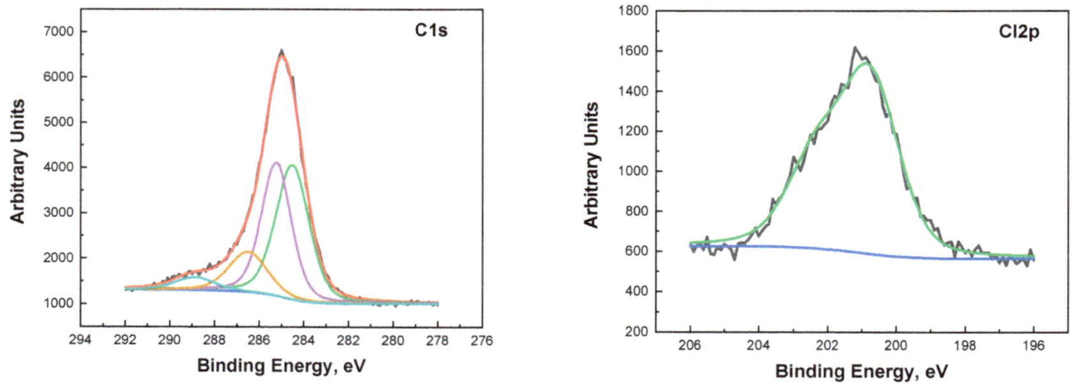

Figure 7. XPS Deconvolution spectrum taken from SPS sample (**a**) C 1s line (**b**), Cl 2p line.

The atomic surface concentrations of the constituent elements in the PS and SPS were calculated and are given in Table 1 and Table 2, respectively.

Table 1. Atomic surface concentrations (%) of the constituent element in the pristine soot sample.

	Pristine Soot				
Concentration (%)	C 1s	O 1s	N 1s	Cl 2p	K 2p
	67.67	26.08	3.36	1.14	1.76

Table 2. Atomic surface concentrations (%) of the constituent elements in the SPS.

	SPS					
Concentration (%)	C 1s	O 1s	N 1s	Cl 2p	Mg 1s	Si 2p
	57.35	20.97	1.19	5.22	0.33	14.93

3.4. Electrochemical Testing

The tested symmetrical cell contained two identical electrodes coated with 80% active material (SPS sample), graphite ABG 1005 EG-1 (10%) and binder PVDF (10%). Binder in powder form and 10 µL N-Methyl-2-pyrrolidone were added to the active SPS mass, and then the prepared paste was applied to a nickel foam current collector with a diameter of 9 mm. The electrodes were dried in a drying oven at 80 °C for 12 h and pressed under pressure (20 MPa).

The cyclic voltammetry measurements (CV) were carried out in a two-electrode cell in a voltage window from 0.05 V up to 1.2 V and different scan rates—from 1 mVs^{-1} up to 50 mVs^{-1}. The CV experiments were carried out on Multi PalmSens model 4. All measurements were performed at RT. The electrochemical test results are shown in Figure 8.

The electrodes were soaked in the electrolyte 6M KOH under vacuum and then mounted in a Swagelok-type cell with Viledon 700/18F separator and filled with 30 µL electrolyte. The capacitor cells were subjected to galvanostatic charge–discharge cycling using an Arbin LBT21084 Instrument System. The results are presented in Figure 9.

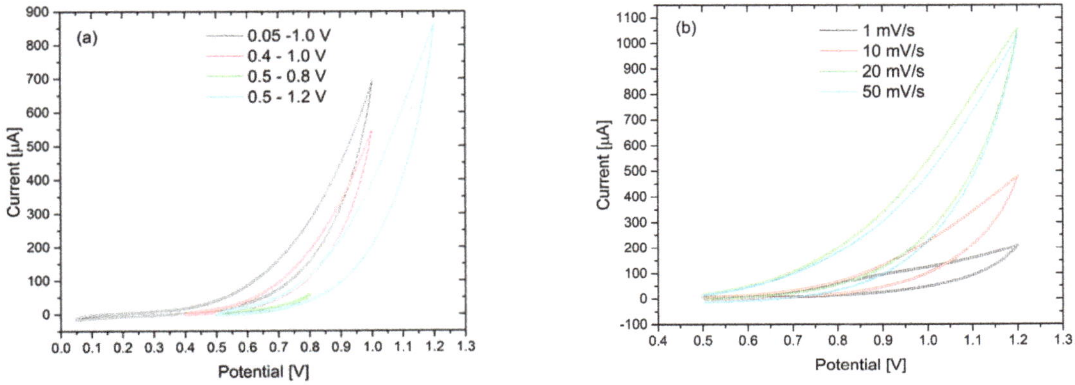

Figure 8. (**a**) CVs within different potential windows at 20 mV s^{-1} scan rate, (**b**) CVs at the scan rates of 1, 10, 20 and 50 mV s^{-1}.

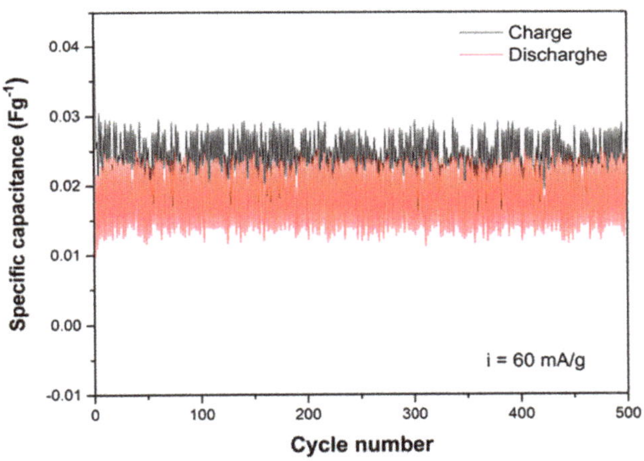

Figure 9. Continuous galvanostatic charge/discharge curves at a current density of 60 mA g^{-1}.

The cyclic voltammetry (CV) studies of the SPS material unveil that an oxidative reaction occurred when applying 6M KOH. There was one noticeable cathodic peak, which appeared at potentials above 0.6–0.8 V (Figure 8a) depending on the voltage window and at rates above 1 mVs^{-1} (Figure 8b). At a rate of 1 mVs^{-1}, the cathodic peak was not as pronounced as can be seen in Figure 8b. Similar behavior at low velocities of the potential unfolding when conducting CV was also observed in our previous studies of thermally treated soot material. [10] The presence of the cathodic peak was due to both the oxygen evolution reaction, which was a consequence of the use of an aqueous electrolyte, and the reduction of chlorine to chloride ion [51]. Due to the unsatisfactory specific electrochemical characteristics of the material investigated in an alkaline electrolyte (6M KOH), more in-depth studies of different electrolyte conditions are planned. Such are envisaged as future work in which to combine the synergistic effect of a chlorine-containing salt in the electrolyte with the possibility of the chlorine ion being intercalated or adsorbed in the carbon matrix.

4. Conclusions

A novel stochastic polymerization procedure was successfully performed on household chimney soot to alter its properties and obtain a material suitable for active electrode coating. The free radical synthesis applied involved a combination of newly created Grignard reagent and chlorinated soot, which created new higher molecular compounds in the original waste product. A Raman spectroscopy investigation performed on the pristine soot and on the final product (SPS) unveiled some significant differences in the carbon structure due to the chemical procedures. The observed decrease in intensity of the G-band in the SPS spectrum compared with the pristine soot spectrum meant a raise in disorder in the graphitic structure. That is connected with the functionalization procedures conducted. The newly created functional groups in the soot did exist in the final product (SPS), which is of high importance for supercapacitance applications. Also, the observed decrease in the background fluorescent signal (FS) in the SPS spectrum compared with the pristine soot spectrum, as well as the complete alignment of the SPS spectrum in the Raman region 2100–2800 cm^{-1}, could be regarded as proof for the removal of any random contamination and obtaining a more uniform final product (SPS). The FTIR spectroscopy investigation showed an increase in the intensity of the peaks, where the C-O, C-C and C=C bending vibrations of alkyl aryl ether, esters and alkenes appeared, meaning new bonds were created in the SPS material. Also, an intensity increase observed for the C-H and C-C spectral features in the FTIR spectrum of the SPS sample is evidence that polymerization was performed correctly, and hydrocarbons with longer chains were created, thus increasing the molecular weight in the final material. The results from the XPS investigation are in agreement with the Raman and FTIR spectroscopy studies. The results showed the appearance of key bonds, such as C-Cl, which is a heritage of the successfully conducted chlorination process. Its appearance is also a piece of evidence that the chlorinated soot and the Grignard agent are in non-stoichiometric quantity. The SPS material was applied as an active electrode coating and was subject to electrochemical testing with KOH as an electrolyte. Further investigations to combine the synergistic effect of a chlorine-containing salt in the electrolyte, with the possibility of the chlorine ion being intercalated or adsorbed in the carbon matrix, should give a positive outcome concerning the electrochemical characteristics of the material.

Author Contributions: Conceptualization, M.P. and T.S.; methodology, M.P.; validation, L.S., D.D., T.M. and I.A.; formal analysis, I.A., T.S., L.S. and T.M.; investigation, B.K., M.P., T.S. and D.D.; resources, M.P., I.A., T.M. and T.S.; writing—original draft preparation, M.P., L.S. and I.A.; writing—review and editing, L.S.; project administration, M.P.; funding acquisition, M.P. All authors have read and agreed to the published version of the manuscript.

Funding: This research was funded by the Bulgarian National Scientific Fund, grant number KP-06-N57/21.

Institutional Review Board Statement: Not applicable.

Informed Consent Statement: Not applicable.

Data Availability Statement: Not applicable.

Acknowledgments: This work was supported by the Bulgarian National Scientific Fund, contract: KP-06-N57/21.

Conflicts of Interest: The authors declare no conflict of interest.

References

1. Anon. Stability of Cycloalkanes—Ring Strain. 2023. Available online: https://chem.libretexts.org/@go/page/31410 (accessed on 20 June 2023).
2. Friebel, F.; Mensah, A.A. Ozone concentration versus temperature: Atmospheric aging of soot particles. *Langmuir* **2019**, *35*, 14437–14450. [CrossRef]
3. Russo, C.; Apicella, B.; Lighty, J.S.; Ciajolo, A.; Tregrossi, A. Optical properties of organic carbon and soot produced in an inverse diffusion flame. *Carbon* **2017**, *124*, 372–379. [CrossRef]

4. Long, C.M.; Nascarella, M.A.; Valberg, P.A. Carbon black vs. black carbon and other airborne materials containing elemental carbon: Physical and chemical distinctions. *Environ. Pollut.* **2013**, *181*, 271–286. [CrossRef] [PubMed]
5. Le, K.C.; Lefumeux, C.; Pino, T. Watching soot inception via online Raman spectroscopy. *Combust. Flame* **2022**, *236*, 111817. [CrossRef]
6. Ray, D.; Bhattacharya, T.S.; Chatterjee, A.; Singha, A.; Ghosh, S.K.; Raha, S. Hygroscopic coating of sulfuric acid shields oxidant attack on the atmospheric pollutant benzo (a) pyrene bound to model soot particles. *Sci. Rep.* **2018**, *8*, 129. [CrossRef]
7. Liu, Y.; He, G.; Chu, B.; Ma, Q.; He, H. Atmospheric heterogeneous reactions on soot: A review. *Fundam. Res.* **2023**, *3*, 579–591. [CrossRef]
8. Sadezky, A.; Muckenhuber, H.; Grothe, H.; Niessner, R.; Pöschl, U. Raman microspectroscopy of soot and related carbonaceous materials: Spectral analysis and structural information. *Carbon* **2005**, *43*, 1731–1742. [CrossRef]
9. Martin, J.W.; Salamanca, M.; Kraft, M. Soot inception: Carbonaceous nanoparticle formation in flames. *Prog. Energy Combust. Sci.* **2022**, *88*, 100956. [CrossRef]
10. Petrov, M.; Lovchinov, K.; Slavov, L.; Stankulov, T.; Nichev, H.; Hikov, T.; Tyutyundzhiev, N. Thermally Stabilized Soot for Supercapacitors. *Phys. Status Solidi A Appl. Res.* **2021**, *218*, 2000617. [CrossRef]
11. Chylek, P.; Jennings, S.G.; Pinnick, R. AEROSOLS | Soot. *Encycl. Atmos. Sci.* **2015**, *2*, 86–91.
12. Lapuerta, M.; Rodríguez–Fernández, J.; Sánchez-Valdepeñas, J. Soot reactivity analysis and implications on diesel filter regeneration. *Prog. Energy Combust. Sci.* **2020**, *78*, 100833. [CrossRef]
13. Sharma, V.; Uy, D.; Gangopadhyay, A.; O'Neill, A.; Paxton, W.A.; Sammut, A.; Ford, M.A.; Aswath, P.B. Structure and chemistry of crankcase and exhaust soot extracted from diesel engines. *Carbon* **2016**, *103*, 327–338. [CrossRef]
14. Yang, Z.; Kornienko, V.; Radchenko, M.; Radchenko, A.; Radchenko, R.; Pavlenko, A. Capture of pollutants from exhaust gases by low-temperature heating surfaces. *Energies* **2022**, *15*, 120. [CrossRef]
15. Ghisellini, P.; Cialani, C.; Ulgiati, S. A review on circular economy: The expected transition to a balanced interplay of environmental and economic systems. *J. Clean. Product.* **2016**, *114*, 11–32. [CrossRef]
16. Bahattin, Y. Osmanli kitap sanatlarinda la'lî mürekkep yapimi. *Süleyman Demirel Üniversitesi İlahiyat Fakültesi Dergisi* **2012**, *28*, 129–142.
17. Uttaravalli, A.N.; Dinda, S.; Kakara, V.R.; Rao, A.R.; Daida, T.; Gidla, B.R. Sustainable use of recycled soot (carbon black) for the cleaner production of value-added products: A compendium. *Chem. Eng. J. Adv.* **2022**, *11*, 100324. [CrossRef]
18. Esmeryan, K.D.; Avramova, I.A.; Castano, C.E.; Ivanova, I.A.; Mohammadi, R.; Radeva, E.I.; Stoyanova, D.S.; Vladkova, T.G. Early stage anti-bioadhesion behavior of superhydrophobic soot based coatings towards Pseudomonas putida. *Mater. Des.* **2018**, *160*, 395–404. [CrossRef]
19. Esmeryan, K.D.; Castano, C.E.; Bressler, A.H.; Abolghasemibizaki, M.; Mohammadi, R. Rapid synthesis of inherently robust and stable superhydrophobic carbon soot coatings. *Appl. Surf. Sci.* **2016**, *369*, 341–347. [CrossRef]
20. Potphode, D.D.; Gangadharan, A.; Sharma, C.S. Carbon soot for electrochemical energy storage applications. *Proc. Indian Natn. Sci. Acad.* **2019**, *85*, 537–551. [CrossRef]
21. Petrov, M.; Lovchinov, K.; Slavov, L.; Hikov, T.; Tyutyundzhiev, N. Electrochemically-stabilized carbon materials for supercapacitor prototypes. *J. Phys. Conf. Ser.* **2021**, *1859*, 012065. [CrossRef]
22. Mao, H.; Li, R.; Huang, C.; Jia, Y.; Wang, W.; Ming, A.; Xiong, J. A highly SERS-active and flexible droplet based on carbon-metal composite nanoparticles. In Proceedings of the 2017 19th International Conference on Solid-State Sensors, Actuators and Microsystems (TRANSDUCERS), Kaohsiung, Taiwan, 18–22 June 2017; pp. 1344–1347. [CrossRef]
23. Colombani, D. Chain-growth control in free radical polymerization. *Prog. Polym. Sci.* **1997**, *22*, 1649–1720. [CrossRef]
24. Geng, K.; Arumugam, V.; Xu, H.; Gao, Y.; Jiang, D. Covalent organic frameworks: Polymer chemistry and functional design. *Prog. Polym. Sci.* **2020**, *108*, 101288. [CrossRef]
25. Braunecker, W.A.; Matyjaszewski, K. Controlled/living radical polymerization: Features, developments, and perspectives. *Prog. Polym. Sci.* **2007**, *32*, 93–146. [CrossRef]
26. Goethals, E.J.; Du Prez, F. Carbocationic polymerizations. *Prog. Polym. Sci.* **2007**, *32*, 220–246. [CrossRef]
27. Rodriguez, K.J.; Pellizzoni, M.M.; Chadwick, R.J.; Guo, C.; Bruns, N. Enzyme-initiated free radical polymerizations of vinyl monomers using horseradish peroxidase. *Methods Enzymol.* **2019**, *627*, 249–262. [CrossRef]
28. Seyferth, D. The Grignard Reagents. *Organometallics* **2009**, *28*, 1598–1605. [CrossRef]
29. Silverman, G.S.; Rakita, P.E. *Handbook of Grignard Reagents*; Marcel Dekker, Inc.: New York, NY, USA, 1996.
30. Yu, G.; Hu, L.; Vosgueritchian, M.; Wang, H.; Xie, X.; McDonough, J.R.; Cui, X.; Cui, Y.; Bao, Z. Solution-processed graphene/MnO_2 nanostructured textiles for high-performance electrochemical capacitors. *Nano Lett.* **2011**, *11*, 2905–2911. [CrossRef] [PubMed]
31. Frackowiak, E.; Beguin, F. Carbon materials for the electrochemical storage of energy in capacitors. *Carbon* **2001**, *39*, 937–950. [CrossRef]
32. Gryglewicz, G.; Machnikowski, J.; Lorenc-Grabowska, E.; Lota, G.; Frackowiak, E. Effect of pore size distribution of coal-based activated carbons on double layer capacitance. *Electrochim. Acta* **2005**, *50*, 1197. [CrossRef]
33. Pandit, S.S.; Juvecar, V.A.; Trivedi, M.K. Stochastic Simulation of polymer reactions. *Chem. Eng. Sci.* **1993**, *48*, 1237. [CrossRef]
34. Feng, D.; Guo, D.; Zhang, Y.; Sun, S.; Zhao, Y.; Shang, Q.; Sun, H.; Wu, J.; Tan, H. Functionalized construction of biochar with hierarchical pore structures and surface O-/N-containing groups for phenol adsorption. *Chem. Eng. J.* **2021**, *410*, 127707. [CrossRef]

35. He, X.; Liu, X.; Nie, B.; Song, D. FTIR and Raman spectroscopy characterization of functional groups in various rank coals. *Fuel* **2017**, *206*, 555–563. [CrossRef]
36. Merlen, A.; Buijnsters, J.G.; Pardanaud, C. A guide to and review of the use of multiwavelength Raman spectroscopy for characterizing defective aromatic carbon solids: From graphene to amorphous carbons. *Coatings* **2017**, *7*, 153. [CrossRef]
37. Pawlyta, M.; Rouzaud, J.N.; Duber, S. Raman microspectroscopy characterization of carbon blacks: Spectral analysis and structural information. *Carbon* **2015**, *84*, 479–490. [CrossRef]
38. Russo, C.; Ciajolo, A. Effect of the flame environment on soot nanostructure inferred by Raman spectroscopy at different excitation wavelengths. *Combust. Flame* **2015**, *162*, 2431–2441. [CrossRef]
39. Ge, H.; Ye, Z.; He, R. Raman spectroscopy of diesel and gasoline engine-out soot using different laser power. *J. Environ. Sci.* **2019**, *79*, 74–80. [CrossRef]
40. Ess, M.N.; Ferry, D.; Kireeva, E.D.; Niessner, R.; Ouf, F.X.; Ivleva, N.P. In situ Raman microspectroscopic analysis of soot samples with different organic carbon content: Structural changes during heating. *Carbon* **2016**, *105*, 572–585. [CrossRef]
41. Zygogianni, A.; Syrigou, M.; Konstandopoulos, A.G.; Kostoglou, M. Oxidative reactivity of particulate samples from different diesel combustion systems and its relation to structural and spectral characteristics of soot. *Emiss. Control Sci. Technol.* **2019**, *5*, 99–123. [CrossRef]
42. Dippel, B.; Jander, H.; Heintzenberg, J. NIR FT Raman spectroscopic study of flame soot. *Phys. Chem. Chem. Phys.* **1999**, *1*, 4707–4712. [CrossRef]
43. Ferrari, A.C.; Robertson, J. Interpretation of Raman spectra of disordered and amorphous carbon. *Phys. Rev. B* **2000**, *61*, 14095–14107. [CrossRef]
44. Baldelli, A.; Trivanovic, U.; Sipkens, T.A.; Rogak, S.N. On determining soot maturity: A review of the role of microscopy-and spectroscopy-based techniques. *Chemosphere* **2020**, *252*, 126532. [CrossRef] [PubMed]
45. Coates, J. Interpretation of infrared spectra, a practical approach. In *Encyclopedia of Analytical Chemistry*; Meyers, R.A., Ed.; John Wiley & Sons, Ltd.: Chichester, UK, 2000; pp. 10815–10837. [CrossRef]
46. Sigma Aldrich Ltd. IR Spectrum Table by Frequency Range. Available online: https://www.sigmaaldrich.com/BG/en/technical-documents/technical-article/analytical-chemistry/photometry-and-reflectometry/ir-spectrum-table (accessed on 2 May 2023).
47. Queiroz, M.F.; Teodosio Melo, K.R.; Sabry, D.A.; Sassaki, G.L.; Rocha, H.A.O. Does the use of chitosan contribute to oxalate kidney stone formation? *Mar. Drugs* **2014**, *13*, 141–158. [CrossRef] [PubMed]
48. Wu, J.; Xie, L.; Li, Y.; Wang, H.; Ouyang, Y.; Guo, J.; Dai, H. Controlled chlorine plasma reaction for noninvasive graphene doping. *J. Am. Chem. Soc.* **2011**, *133*, 19668–19671. [CrossRef] [PubMed]
49. Li, B.; Zhou, L.; Wu, D.; Peng, H.; Yan, K.; Zhou, Y.; Liu, Z. Photochemical chlorination of graphene. *ACS Nano* **2011**, *5*, 5957–5961. [CrossRef]
50. Gopalakrishnan, K.; Subrahmanyam, K.S.; Kumar, P.; Govindaraj, A.; Rao, C.N.R. Reversible chemical storage of halogens in few-layer graphene. *RSC Adv.* **2012**, *2*, 1605–1608. [CrossRef]
51. Liang, G.; Liang, B.; Chen, A.; Zhu, J.; Li, Q.; Huang, Z.; Li, X.; Wang, Y.; Wang, X.; Xiong, B.; et al. Development of rechargeable high-energy hybrid zinc-iodine aqueous batteries exploiting reversible chlorine-based redox reaction. *Nat. Commun.* **2023**, *14*, 1856. [CrossRef]

coatings

MDPI

Article

Effect of Deposition Parameters on Morphological and Compositional Characteristics of Electrodeposited $CuFeO_2$ Film

Min-Kyu Son

Nano Convergence Materials Center, Emerging Materials R&D Division, Korea Institute of Ceramic Engineering & Technology (KICET), Jinju 52851, Republic of Korea; minkyu.son@kicet.re.kr; Tel.: +82-55-792-2683

Abstract: Deposition parameters determine the characteristics of semiconductor films in electrodeposition. Thus, it is essential to understand the effect of deposition parameters on the electrodeposited film for fabricating suitable semiconductor film fitting for various applications. In this work, the morphological and compositional properties of electrodeposited delafossite $CuFeO_2$ film, according to the deposition parameters, were studied. The $CuFeO_2$ film was fabricated by the galvanostatic electrodeposition and post-annealing process under inert gas flow. The type of solvent, electrolyte condition, applied current density and deposition time were controlled as the variable deposition parameters. As a result, the typical $CuFeO_2$ film, without any impurities, was electrodeposited in the electrolyte-based DMSO solvent. Interestingly, the concentration of potassium perchlorate as a complexing agent caused morphological change in electrodeposited $CuFeO_2$ film, as well as compositional transition. On the other hand, the applied current density and deposition time only influenced the morphology of electrodeposited $CuFeO_2$ film. These observations would provide specific guidelines for the fabrication of electrodeposited $CuFeO_2$ film with suitable composition and morphology for various applications.

Keywords: delafossite $CuFeO_2$ film; galvanostatic electrodeposition; deposition parameters; morphological characteristic; compositional characteristic

Citation: Son, M.-K. Effect of Deposition Parameters on Morphological and Compositional Characteristics of Electrodeposited $CuFeO_2$ Film. *Coatings* **2022**, *12*, 1820. https://doi.org/10.3390/coatings12121820

Academic Editor: Heping Li

Received: 30 October 2022
Accepted: 22 November 2022
Published: 25 November 2022

1. Introduction

Delafossite-structured $CuFeO_2$ has attracted much interest in various research fields due to its inherent characteristics, as shown in Table 1. It has been widely used as a transparent conductive film owing to its high p-type conductivity [1–3]. It also shows a relatively high Seebeck coefficient, which is beneficial in temperature sensor applications [4,5]. In addition, it exhibits a unique magnetic behavior, inducing phase transition at low temperature by antiferromagnetic interactions between Fe^{3+} ions in $CuFeO_2$. Hence, it can be utilized in multifunctional magnetoelectric devices [6,7]. Contrary to other wide band gap delafossite materials, its band gap is visibly light-responsive (1.1~1.6 eV) [8,9]. Therefore, in recent years, it has been extensively investigated in solar energy conversion devices, such as solar cells [10,11], photoelectrochemical photocathodes for water splitting [9,12–16] and photocatalysts [17–19].

Table 1. General characteristics of delafossite-structured $CuFeO_2$ [1–3,8,9,16].

Characteristics	Values
Conductivity	1.53~2 S cm^{-1}
Carrier mobility	0.2 $cm^2\ V^{-1}\ s^{-1}$
Hall coefficient	$1.84 \times 10^6\ m^2\ C^{-1}$
Band gap	1.1~1.6 eV
Absorption coefficient	Up to $10^7\ m^{-1}$

$CuFeO_2$ thin film has been considered as a one-size-fits-all electrode in these application devices. It has been synthesized via various deposition techniques such as sputtering [2,20–22], pulsed laser deposition [23–25], sol-gel based spin coating [1,26,27], hydrothermal method [19,28], electrodeposition [9,29–31] and spray pyrolysis [3,32,33]. Among these deposition techniques, the electrodeposition is an advantageous method for fabrication of high-quality $CuFeO_2$ thin film. First, it is cost-effective because it does not need expensive vacuum facilities, contrary to sputtering or pulsed laser deposition. The only necessary tools for electrodeposition are an electrolyte bath and a potentiostat to apply the potential or current. Second, it is an energy-saving deposition method because the process temperature is mild (room temperature or below 100 °C), unlike the hydrothermal method or spray pyrolysis. Third, it is easily scalable in a large area bath, facilitating the fabrication of large area $CuFeO_2$ electrodes [34,35].

Read et al. successfully fabricated an electrodeposited $CuFeO_2$ photocathode using a deposition solution based on dimethyl sulfoxide (DMSO) solvent [9]. This deposition solution also includes 1 mM copper (II) nitrate, 3 mM iron (III) perchlorate, and 100 mM potassium perchlorate. Crystalline $CuFeO_2$ film with a thickness of 100~130 nm was fabricated by the potentiostatic mode for 20 min and the sequential annealing process under inert gas atmosphere. Riveros et al. also obtained $CuFeO_2$ thin film from DMSO based deposition solution [36]. They studied the characteristics of electrodeposited $CuFeO_2$ film by controlling the potential and the anion type in the deposition solution. As a result, stoichiometric $CuFeO_2$ film was grown by the electrodeposition with an applied potential of −0.6 V in perchlorate/chloride anions-mixed DMSO solution. Kang et al. fabricated $CuFeO_2$ and CuO composite films by potentiostatic electrodeposition using aqueous solution containing 4 mM copper (II) nitrate, 12 mM iron (III) perchlorate and 50 mM potassium perchlorate [37]. It was revealed that the morphology of a deposited $CuFeO_2$/CuO film based water solution is quite different when compared with the DMSO solution. The whole electrodeposition process is almost identical in the relevant literature. Nevertheless, the detailed electrodeposition conditions are slightly different. In electrodeposition, deposition parameters such as potential, current density, time and electrolyte composition determine the characteristics of the $CuFeO_2$ film [38]. Therefore, it is important to control these during the electrodeposition process to obtain the desired $CuFeO_2$ film in accordance with the specific application. Furthermore, it is essential to study the effect of deposition parameters on the characteristics of electrodeposited $CuFeO_2$ film.

Thus, in this study, the morphological and compositional characteristics of electrodeposited $CuFeO_2$ film, depending on these conditions, were investigated. Two types of solvent were selected: DMSO and water, which are typical solvents for electrodepositing $CuFeO_2$ film. The concentration of chemical salts in the electrolyte was chosen from representative recipes in previous literature [9,36,37]. The $CuFeO_2$ film was electrodeposited by galvanostatic mode, enabling precise thickness control. The current and deposition time were also controlled because they are significant deposition parameters that determine the property of an electrodeposited film in galvanostatic electrodeposition. Herein, the understanding of the characteristics of the electrodeposited $CuFeO_2$ film, depending on these electrodeposition parameters, would provide guidelines for selecting suitable deposition conditions to fabricate the optimal electrodeposited $CuFeO_2$ electrode fitting for specific applications.

2. Materials and Methods

DMSO (C_2H_6SO, 99.9%, Sigma Aldrich, St. Louis, MO, USA) and distilled water were used as the solvent to prepare the electrolyte for electrodeposition. To examine the solvent effect, electrolyte containing 1 mM copper (II) nitrate hydrate ($Cu(NO_3)_2 \cdot xH_2O$, 99.999%, Sigma Aldrich, St. Louis, MO, USA), 3 mM iron (III) perchlorate hydrate ($Fe(ClO_4)_3 \cdot xH_2O$, Sigma Aldrich, St. Louis, MO, USA), and 100 mM potassium perchlorate ($KClO_4$, 99%, Sigma Aldrich, St. Louis, MO, USA) was prepared. On the other hand, electrolyte with more concentration of Cu/Fe salts and less concentration of $KClO_4$ was prepared to study the effect of electrolyte composition on $CuFeO_2$ film. It contained 4 mM $Cu(NO_3)_2{\cdot}xH_2O$, 12 mM $Fe(ClO_4)_3{\cdot}xH_2O$ and 50 mM potassium perchlorate. The ratio of Cu and Fe was fixed to 1:3 in all electrolytes for electrodeposition.

The $CuFeO_2$ film was electrodeposited on a cleaned fluorine-doped tin oxide (FTO) glass substrate (surface resistivity 7 Ω sq^{-1}, Sigma Aldrich, St. Louis, MO, USA). The cleaning process of FTO glass substrates consisted of ultrasonication for 30 min then O_2 plasma treatment (CUTE, FEMTO Science, Hwaseong, Republic of Korea) for 10 min. The ultrasonication was carried out in the order of acetone, ethanol, and distilled water and each process was carried out for 10 min. The electrodeposition was carried out in the prepared electrolyte via the galvanostatic mode using a standard three electrode system. It was composed of the FTO glass substrate as a working electrode, the Pt wire as a counter electrode and the Ag/AgCl electrode in saturated KCl as a reference electrode. Constant current density was applied by a potentiostat (HSV-100, Hokuto Denko, Tokyo, Japan) during electrodeposition at room temperature for various deposition time periods. Finally, the electrodeposited $CuFeO_2$ film was annealed at 650 °C for 1 h with a ramp ratio of 5 °C min^{-1} under inert N_2 gas flow.

The morphological characteristics of fabricated $CuFeO_2$ films was analyzed by a high-resolution scanning electron microscope (SEM, JSM-7900F, JEOL Ltd., Tokyo, Japan). In the SEM analyses, the accelerating voltage was 5.0 kV, while the working distance was 10.0 mm. The element analysis on the $CuFeO_2$ film surface was conducted by an energy dispersive X-ray (EDX, Oxford Instrument, Abingdon-on-Thames, UK) analyzer attached to the SEM system. The crystallographic analyses were performed using a high-resolution X-ray diffraction system (XRD, SmartLab 9 kW AMK, Rigaku Corporation, Tokyo, Japan) and an Arc cluster ion beam X-ray photoelectron spectroscopy system (XPS, PHI 5000 Versa Probe II, ULVAC, Kanagawa, Japan). The XRD measurement was carried out in the range of 2θ = 20° to 70° with a step width of 0.02° and a scan rate of 2° min^{-1} using a Cu-Kα radiation source. Measured XPS spectra were fitted by the PHI Multipak software.

3. Results and Discussion

Non-aqueous based electrodeposition does not form metal hydroxide during the electrodeposition process, contrary to aqueous based electrodeposition. Thus, it influences the property of electrodeposited thin film. To examine this influence, the $CuFeO_2$ film was electrodeposited using DMSO or distilled water solution with 1 mM $Cu(NO_3)_2{\cdot}xH_2O$, 3 mM $Fe(ClO_4)_3{\cdot}xH_2O$ and 100 mM potassium perchlorate. Figure 1 shows the XRD patterns of the electrodeposited film in the electrolyte based DMSO and water solvents by applying a current density of −0.2 mA cm^{-2} for 60 min with post-annealing treatment at 650 °C for 1 h under N_2 gas flow. All samples have diffraction peaks related to SnO_2 (JCPDS No. 46-1088) from the FTO glass substrate. The diffraction peaks at 2θ = 31.26° and 35.8°, attributed to (006) and (012) orientations of the crystalline $CuFeO_2$ (JCPDS No. 0175-2146) [39–41], were observed in the electrodeposited film in the DMSO solution. No other diffraction peaks, including those of metallic Cu or Fe, were observed. This means that electrodeposition using a DMSO solvent and post-annealing process produces a well-crystalline $CuFeO_2$ film without any impurities. On the other hand, diffraction peaks at 2θ = 36.4° and 42.3°, corresponding to (111) and (200) orientations of the crystalline Cu_2O (JCPDS No. 05-0667) [42,43], respectively, were observed in the electrodeposited film in the water solution. This indicates that the electrodeposited film in the water solution is the Cu_2O film and not the $CuFeO_2$ film.

The reason for this difference can be found in the mechanism of electrodeposition in different solvents. In general, the electrodeposited $CuFeO_2$ film was formed by co-deposition of Cu_2O and Fe_2O_3 via the following reactions [36]:

$$Fe^{3+}{}_{(sol)} + e^- \rightarrow Fe^{2+}{}_{(sol)} \quad (1)$$

$$Cu^{2+}{}_{(sol)} + e^- \rightarrow Cu^{+}{}_{(sol)} \quad (2)$$

$$2\,Fe^{2+}{}_{(sol)} + 3/2\,O_{2(sol)} + 4\,e^- \rightarrow Fe_2O_{3(s)} \quad (3)$$

$$2\,Cu^{+}{}_{(sol)} + 1/2\,O_{2(sol)} + 2\,e^- \rightarrow Cu_2O_{(s)} \quad (4)$$

$$Cu_2O_{(s)} + Fe_2O_{3(s)} \rightarrow 2\,CuFeO_{2(s)} \quad (5)$$

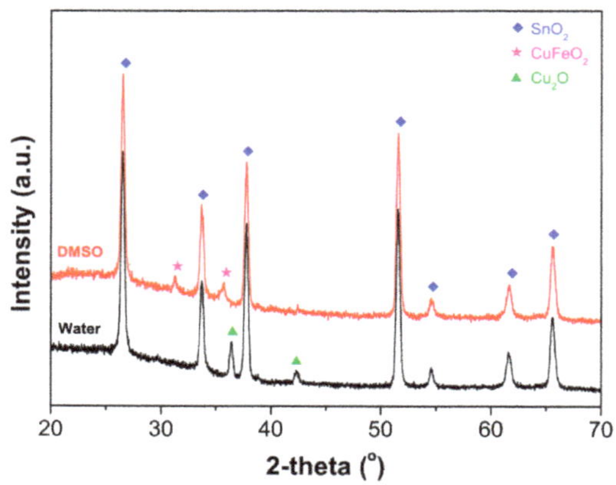

Figure 1. XRD patterns of electrodeposited film in electrolyte-based DMSO (red) and water (black) solvents. Electrodeposition was carried out by applying a current density of −0.2 mA cm^{-2} for 60 min and the films were annealed at 650 °C for 1 h under N_2 gas flow after electrodeposition.

In these reactions, the formation of Cu_2O is more favored than that of Fe_2O_3 because the former is kinetically preferred to the latter [36]. Hence, the Cu_2O film was formed in the electrodeposition using the water solution. Meanwhile, molecules of DMSO ($(CH_3)_2SO$) form strong complexes with copper (II) ions in the copper (II) nitrate hydrate [44]. This stabilizes the Cu ions in the electrodeposition bath, accelerating the Fe deposition in the film. Therefore, the $CuFeO_2$ film was formed in the electrodeposition using DMSO solution.

The pristine electrodeposited $CuFeO_2$ film from the DMSO solution was amorphous because no diffraction peaks were detected, except those related to the SnO_2 from the FTO substrate (Figure S1). This means that the post-annealing treatment at 650 °C under N_2 gas flow is necessary to transform the amorphous $CuFeO_2$ into the crystalline one. In addition, it is clearly shown that the crystalline $CuFeO_2$ film was homogeneously distributed, by element mapping of Cu, Fe and O from the top-view EDX characterization (Figure S2). Moreover, from the EDX characterization, it was demonstrated that the atomic ratio of Cu:Fe is almost 1:1 (9.87:8.88), which also indicates that the electrodeposited $CuFeO_2$ film is stoichiometric. This suggests that the homogeneous and stoichiometric crystalline $CuFeO_2$ film is successfully fabricated by the electrodeposition using a DMSO-based solution and post-annealing process under inert gas flow.

A complexing agent in the deposition electrolyte has been introduced to activate the Fe deposition for the electrodeposited $CuFeO_2$ film. Therefore, it could have affected the characteristic of electrodeposited $CuFeO_2$ film. Potassium perchlorate has been used to provide the chloride anion as a complexing agent in the electrodeposition of $CuFeO_2$ film [9,36,37]. To investigate the effect of the complexing agent on the characteristics of electrodeposited $CuFeO_2$ film, the electrodeposition was carried out using electrolyte with different potassium perchlorate concentrations. Figure 2 exhibits the XRD patterns of electrodeposited $CuFeO_2$ film by applying a current density of −0.2 mA cm^{-2} for 60 min in the DMSO based electrolyte containing 1 mM $Cu(NO_3)_2{\cdot}xH_2O$/3 mM $Fe(ClO_4)_3{\cdot}xH_2O$/100 mM potassium perchlorate (Solution #1) and 4 mM $Cu(NO_3)_2{\cdot}xH_2O$/12 mM $Fe(ClO_4)_3{\cdot}xH_2O$/50 mM potassium perchlorate (Solution #2) after the post-annealing treatment at 650 °C under N_2 gas flow. The diffraction peaks at 2θ = 31.26° and 35.8° are well matched with (006) and (012) orientations of the crystalline $CuFeO_2$ in the electrodeposited film in Solution

#1. On the other hand, in the electrodeposited film in Solution #2, the diffraction peak corresponding to the (006) orientation of $CuFeO_2$ was also observed. However, the diffraction peak related to the (012) orientation of $CuFeO_2$ was slightly shifted to the large angle (35.8° → 36.26°). It is assumed that the characteristic of $CuFeO_2$ film was affected by the lattice parameter change or defect density [45]. In addition, the diffraction peak ($2\theta = 42.3°$) attributed to the (111) orientation of the crystalline Cu_2O was observed. This indicates that Cu_2O was formed as the impurity when the film was electrodeposited in Solution #2.

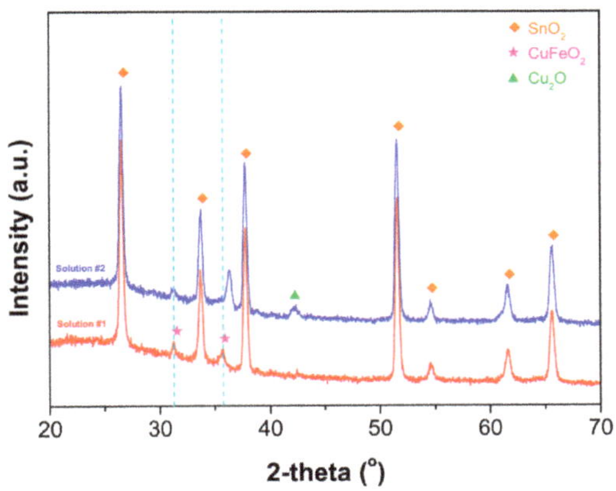

Figure 2. XRD patterns of electrodeposited $CuFeO_2$ film by applying a current density of −0.2 mA cm^{-2} for 60 min in DMSO based electrolytes with different potassium perchlorate concentrations. Solution #1 (red) contained 1 mM $Cu(NO_3)_2 \cdot xH_2O$/3 mM $Fe(ClO_4)_3 \cdot xH_2O$/100 mM potassium perchlorate, while Solution #2 (blue) contained 4 mM $Cu(NO_3)_2 \cdot xH_2O$/12 mM $Fe(ClO_4)_3 \cdot xH_2O$/50 mM potassium perchlorate. Electrodeposited films were annealed at 650 °C for 1 h under N_2 gas flow.

Besides this, the morphologies of two films were totally different. As shown in Figure 3, irregularly shaped particles were covered in the electrodeposited film in Solution #1 (Figure 3a), while spherically shaped particles were deposited in the electrodeposited film in Solution #2 (Figure 3b). The average particle size was approximately 220 nm and some particles were agglomerated in the electrodeposited film in Solution #1. Meanwhile, the average particle size was approximately 315 nm and no particle aggregations were observed in the electrodeposited film in Solution #2. Based on these observations, it was concluded that the electrolyte containing more potassium perchlorate is adequate for making the pure crystalline $CuFeO_2$ film with small particles using the galvanostatic electrodeposition. It is also demonstrated that the concentration of Cu/Fe salts was not a dominant factor determining the characteristic of the electrodeposited $CuFeO_2$ film in the electrolyte condition, with a fixed ratio of Cu:Fe = 1:3.

The applied current density was controlled during the electrodeposition in the DMSO based electrolyte containing 1 mM $Cu(NO_3)_2 \cdot xH_2O$/3 mM $Fe(ClO_4)_3 \cdot xH_2O$/100 mM potassium perchlorate to study the property of the electrodeposited $CuFeO_2$ film according to the current density in the galvanostatic mode. Figure 4a show the XRD patterns of the electrodeposited $CuFeO_2$ film for 30 min by applying different current densities after post-annealing treatment at 650 °C under N_2 gas flow. As illustrated in Figure 4a, all films have diffraction peaks at $2\theta = 31.26°$ and 35.8°, attributed to the (006) and (012) orientations of the crystalline $CuFeO_2$. Although they are weaker than those in the $CuFeO_2$ electrodeposited for 60 min, due to the short deposition time, it clearly supports that the deposited film was the crystalline $CuFeO_2$ without any impurities. In other words, the current density of galvanostatic electrodeposition did not have any influence on the composition of the electrodeposited film. However, it affected the morphology of the electrodeposited $CuFeO_2$ film, as shown in Figure 4b–d. The irregular shaped $CuFeO_2$ particles were formed on the substrate, similar to that in Figure 3a, while the aggregation of $CuFeO_2$ particles became large when the applied current density was increased. This was likely to be due to the distinction of deposition speed, depending on the current density. The Cu/Fe ions in the solution were quickly moved to the substrate's surface or pre-deposited $CuFeO_2$ film by the strong electric field when the large current density was applied. Hence, the aggregation became severe around the previously deposited $CuFeO_2$ particles in the electrodeposited film with the large current density.

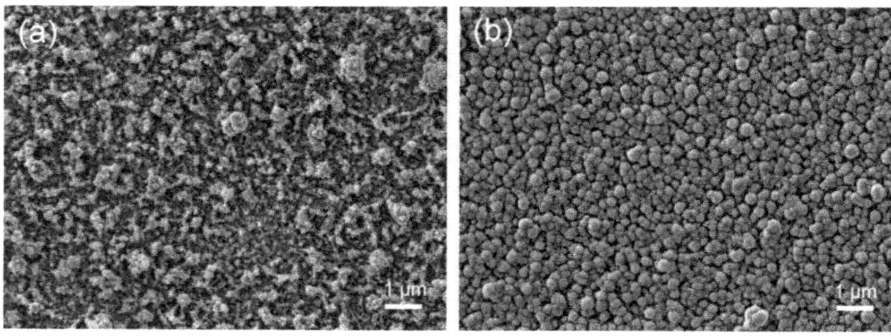

Figure 3. Top-view SEM images of electrodeposited $CuFeO_2$ films when applying a current density of -0.2 mA cm^{-2} for 60 min in DMSO based electrolytes with different potassium perchlorate concentrations: (**a**) Solution #1 containing 1 mM $Cu(NO_3)_2{\cdot}xH_2O$/3 mM $Fe(ClO_4)_3{\cdot}xH_2O$/100 mM potassium perchlorate and (**b**) Solution #2 containing 4 mM $Cu(NO_3)_2{\cdot}xH_2O$/12 mM $Fe(ClO_4)_3{\cdot}xH_2O$/50 mM potassium perchlorate. Electrodeposited films were annealed at 650 °C for 1 h under N_2 gas flow.

Deposition time is a main factor in controlling the thickness of an electrodeposited film. Figure 5 shows the cross-section SEM images of electrodeposited $CuFeO_2$ film with an applied current density of -0.1 mA cm^{-2} in the DMSO based electrolyte containing 1 mM $Cu(NO_3)_2{\cdot}xH_2O$/3 mM $Fe(ClO_4)_3{\cdot}xH_2O$/100 mM potassium perchlorate after post-annealing treatment at 650 °C under N_2 gas flow. The electrodeposition for 20 min produced a homogeneous nanostructured $CuFeO_2$ film with an average thickness of 500 nm, as shown in Figure 5a. The average thickness was increased to 875 nm after the electrodeposition for 40 min, as illustrated in Figure 5b. Finally, the nanostructured $CuFeO_2$ film with an average thickness of 1000 nm was electrodeposited for 60 min (Figure 5c). Interestingly, the surface of nanostructured $CuFeO_2$ film became rougher as the deposition time increased. In addition, the deposition of $CuFeO_2$ film became slow as the deposition time passed. In other words, it means that the deposition ratio over time was not linear. The rougher nanostructured $CuFeO_2$ film with the long deposition time was mainly due to the aggregation of $CuFeO_2$ particles, while the non-linear deposition ratio over time was likely to be due to the relatively low conductivity of the electrode by the pre-deposited $CuFeO_2$ film on the substrate.

In this way, it was demonstrated that electrodeposition in the DMSO based electrolyte containing more potassium perchlorate concentration with an applied current density of -0.1 mA cm^{-2} for deposition time below 30 min, and the post-annealing process at 650 °C under inert gas flow, produces nanostructured $CuFeO_2$ film with less aggregations. The XPS analysis also confirms the composition of this film in detail. Figure 6 shows the XPS spectra of the electrodeposited $CuFeO_2$ film in the DMSO based electrolyte containing 1 mM $Cu(NO_3)_2{\cdot}xH_2O$/3 mM $Fe(ClO_4)_3{\cdot}xH_2O$/100 mM potassium perchlorate with an applied current density of -0.1 mA cm^{-2} for 30 min after post-annealing treatment at 650 °C under N_2 gas flow: Cu 2p (Figure 6a), Fe 2p (Figure 6b), and O 1s (Figure 6c). In Figure 6a, peaks located at 932 eV and 952 eV correspond to the binding energies of Cu (I) $2p_{3/2}$ and Cu (I) $2p_{1/2}$, respectively. This confirms the monovalent state of Cu (Cu^+) in the $CuFeO_2$ film [46,47]. In Figure 6b, peaks located at 710 eV and 723 eV are derived from the binding energies of Fe (III) $2p_{3/2}$ and Fe (III) $2p_{1/2}$, indicating the trivalent state of Fe (Fe^{3+}) in the $CuFeO_2$ film [46,47]. As shown in Figure 6c, the O 1s spectrum was deconvoluted into two peaks located at 529 eV and 530 eV, corresponding to the lattice oxygen species [48,49]. This is in good agreement with the typical chemical status of the crystalline $CuFeO_2$ in the literature [36]. Based on these compositional and morphological analyses, it was concluded that the DMSO based solution with more complexing agent concentrations is a suitable electrolyte for the electrodeposition bath to fabricate the nanostructured $CuFeO_2$ film without any impurities. In addition, it was demonstrated that it is possible to control the morphology of nanostructured $CuFeO_2$ film without any compositional transitions by changing the current density and deposition time of galvanostatic electrodeposition.

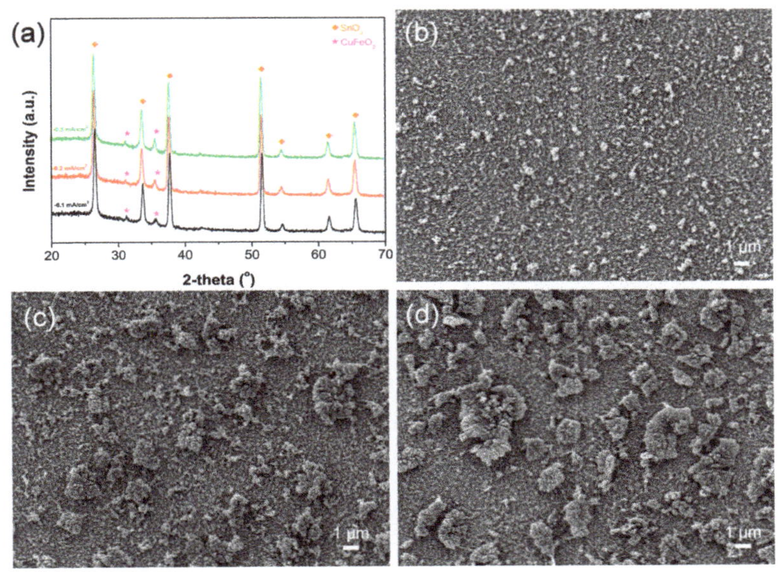

Figure 4. (**a**) XRD patterns of electrodeposited $CuFeO_2$ film by applying different current densities: -0.1 mA cm^{-2} (black), -0.2 mA cm^{-2} (red), and -0.3 mA cm^{-2} (green). Top view SEM images of electrodeposited $CuFeO_2$ film by applying different current densities of (**b**) -0.1 mA cm^{-2}, (**c**) -0.2 mA cm^{-2}, and (**d**) -0.3 mA cm^{-2}. Electrodeposition was carried out in DMSO based electrolytes containing 1 mM $Cu(NO_3)_2{\cdot}xH_2O$/3 mM $Fe(ClO_4)_3{\cdot}xH_2O$/100 mM potassium perchlorate for 30 min. Electrodeposited films were annealed at 650 °C for 1 h under N_2 gas flow.

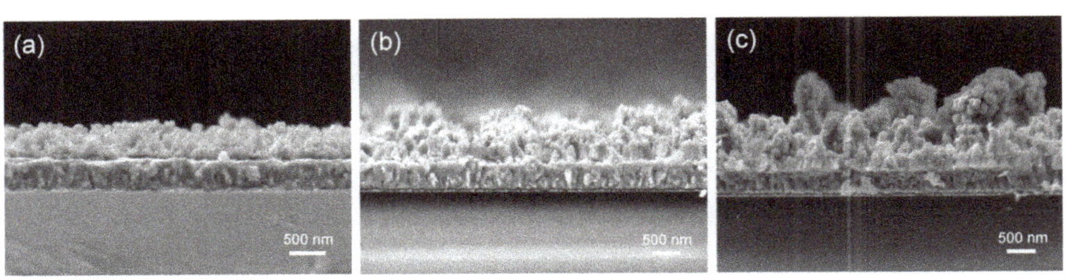

Figure 5. Cross-section SEM images of electrodeposited $CuFeO_2$ film with an applied current density of -0.1 mA cm^{-2} in DMSO based electrolyte containing 1 mM $Cu(NO_3)_2{\cdot}xH_2O$/3 mM $Fe(ClO_4)_3{\cdot}xH_2O$/100 mM potassium perchlorate for different deposition time: (**a**) 20 min (**b**) 40 min and (**c**) 60 min. Samples were annealed at 650 °C for 1 h under N_2 gas flow after electrodeposition.

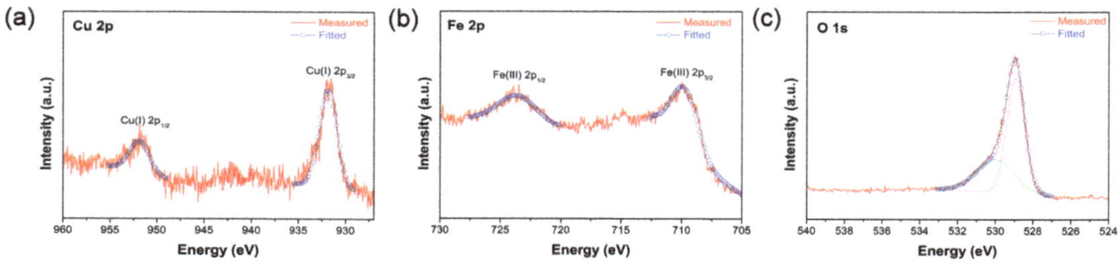

Figure 6. XPS spectra of electrodeposited $CuFeO_2$ film in DMSO based electrolyte containing 1 mM $Cu(NO_3)_2 \cdot xH_2O$/3 mM $Fe(ClO_4)_3 \cdot xH_2O$/100 mM potassium perchlorate with an applied current density of -0.1 mA cm^{-2} for 30 min after post-annealing treatment at 650 °C for 1hr under N_2 gas flow: (**a**) Cu 2p, (**b**) Fe 2p, and (**c**) O 1s.

4. Conclusions

In this study, the effect of deposition parameters, including the type of solvent, the condition of electrolyte, the applied current density and deposition time. on morphological and compositional characteristics of electrodeposited $CuFeO_2$ film was investigated. As a result, in terms of the solvent, nanostructured $CuFeO_2$ film was fabricated using the DMSO solution, while Cu_2O film was formed using the water solution. Furthermore, the concentration of potassium perchlorate as a complexing agent in the electrolyte caused morphological change in the electrodeposited $CuFeO_2$ film, as well as the compositional transition. On the other hand, the applied current density and deposition time did not have an influence on the composition of the electrodeposited $CuFeO_2$ film. However, they caused the morphological changes in the electrodeposited $CuFeO_2$ film. Along with previous studies (Table S1), it is expected that this will provide guidelines for selecting suitable electrodeposition conditions to fabricate nanostructured $CuFeO_2$ composite electrode for specific applications, such as solar energy conversion devices, temperature sensors, photocatalysts, magnetoelectric devices and transparent conductive substrates.

Supplementary Materials: The following supporting information can be downloaded at: https://www.mdpi.com/article/10.3390/coatings12121820/s1, Figure S1: The XRD pattern of pristine electrodeposited $CuFeO_2$ film in the DMSO-based electrolyte containing 1 mM $Cu(NO_3)_2 \cdot xH_2O$, 3 mM $Fe(ClO_4)_3 \cdot xH_2O$, and 100 mM potassium perchlorate by applying a current density of -0.2 mA cm^{-2} for 60 min; Figure S2: The top-view EDX characterization of the electrodeposited $CuFeO_2$ film in the DMSO-based electrolyte containing 1 mM $Cu(NO_3)_2 \cdot xH_2O$, 3 mM $Fe(ClO_4)_3 \cdot xH_2O$, and 100 mM potassium perchlorate by applying a current density of -0.2 mA cm^{-2} for 60 min after post annealing treatment at 650 °C for 60 min under N_2 gas flow; Table S1: Comparison of electrodeposition conditions and characteristics of electrodeposited film between previous studies and this work.

Funding: This research was supported in part by the National Research Foundation of Korea (NRF), grant funded by the Korean government (MSIT) (NRF-2021R1F1A1059126) and in part by the program of Future Hydrogen Original Technology Development (NRF-2021M3I3A1084649) through the National Research Foundation of Korea (NRF), funded by the Korean government (MSIT).

Institutional Review Board Statement: Not applicable.

Informed Consent Statement: Not applicable.

Data Availability Statement: The data presented in this study are available on request from the corresponding author.

Conflicts of Interest: The authors declare no conflict of interest.

References

1. Chen, H.-Y.; Wu, J.-H. Transparent conductive $CuFeO_2$ thin films prepared by sol-gel processing. *Appl. Surf. Sci.* **2012**, *258*, 4844–4847. [CrossRef]
2. Deng, Z.; Fang, X.; Wang, X.; Wu, S.; Dong, W.; Shao, J.; Tao, R. Characterization of amorphous p-type transparent $CuFeO_2$ thin films prepared by radio frequency magnetron sputtering method at room temperature. *Thin Solid Film.* **2015**, *589*, 17–21. [CrossRef]
3. Mohamed, H.; Chikoidze, E.; Ratep, A.; Elsoud, A.M.A.; Boshta, M.; Osman, M.B.S. Synthesis of conducting single-phase $CuFeO_2$ thin films by spray pyrolysis technique. *Mater. Sci. Semicond. Process* **2020**, *107*, 104831. [CrossRef]
4. Elgazzar, E.; Tataroglu, A.; Al-Ghamdi, A.A.; Al-Tuki, Y.; Farooq, W.A.; El-Tantawy, F.; Yakuphanoglu, F. Thermal sensors based on delafossite film/p-silicon diode for low-temperature measurements. *Appl. Phys. A* **2016**, *122*, 617. [CrossRef]
5. Sinnarasa, I.; Thimont, Y.; Barnabe, A.; Beaudhuin, M.; Moll, A.; Schorne-Pinto, J.; Tailhades, P.; Presmanes, L. Microstructural and transport properties of Mg-doped $CuFeO_2$ thin films: A promising material for high accuracy miniaturized temperature sensors based on the Seebeck effect. *J. Alloys Compd.* **2020**, *827*, 154199. [CrossRef]
6. Xia, N.; Shi, L.; Xia, Z.; Chen, B.; Jin, Z.; Wang, Y.; Ouyang, Z.; Zuo, H.; Shen, Y. Dynamic behavior of magnetoelectric coupling of $CuFeO_2$ induced by a high magnetic field. *J. Appl. Phys.* **2014**, *115*, 114107. [CrossRef]
7. Dai, H.; Ye, F.; Li, T.; Chen, Z.; Cao, X.; Wang, B. Impact of Li doping on the microstructure, defects, and physical properties of $CuFeO_2$ multiferroic ceramics. *Ceram. Int.* **2019**, *45*, 24570–24577. [CrossRef]
8. Crespo, C.T. Potentiality of $CuFeO_2$-delafossite as a solar energy converter. *Sol. Energy* **2018**, *163*, 162–166. [CrossRef]
9. Read, C.G.; Park, Y.; Choi, K.-S. Electrochemical synthesis of p-type $CuFeO_2$ electrodes for use in a photoelectrochemical cell. *J. Phys. Chem. Lett.* **2012**, *3*, 1872–1876. [CrossRef]
10. Zhu, T.; Deng, Z.; Fang, X.; Huo, Z.; Wang, S.; Dong, W.; Shao, J.; Tao, R.; Song, C.; Wang, L. High photovoltages of $CuFeO_2$ based p-type dye-sensitized solar cells. *J. Alloys Compd.* **2016**, *685*, 836–840. [CrossRef]
11. Jin, Y.; Chumanov, G. Solution synthesis of pure 2H $CuFeO_2$ at low temperature. *RSC Adv.* **2016**, *6*, 26392–26397. [CrossRef]
12. Prevot, M.S.; Guijarro, N.; Sivula, K. Enhancing the performance of a robust sol-gel processed-type delafossite $CuFeO_2$ photocathode for solar water reduction. *ChemSusChem* **2015**, *8*, 1359–1367. [CrossRef]
13. Jang, Y.J.; Park, Y.B.; Kim, H.E.; Choi, Y.H.; Choi, S.H.; Lee, J.S. Oxygen-intercalated $CuFeO_2$ photocathode fabricated by hybrid microwave annealing for efficient solar hydrogen production. *Chem. Mater.* **2016**, *28*, 6054–6061. [CrossRef]
14. Oh, Y.; Yang, W.; Kim, J.; Jeong, S.; Moon, J. Enhanced photocurrent of transparent $CuFeO_2$ photocathodes by self-light-harvesting architecture. *ACS Appl. Mater. Interfaces* **2017**, *9*, 14078–14087. [CrossRef]
15. Boudoire, F.; Liu, Y.; Formal, F.L.; Guijarro, N.; Lhermitte, C.R.; Sivula, K. Spray synthesis of $CuFeO_2$ photocathodes and in-operando assessment of charge carrier recombination. *J. Phys. Chem. C* **2021**, *125*, 10883–10890. [CrossRef]
16. Prevot, M.S.; Jeanbourquin, X.A.; Bouree, W.S.; Abdi, F.; Friedrich, D.; Krol, R.; Guijarro, N.; Formal, F.L.; Sivula, K. Evaluating charge carrier transport and surface states in $CuFeO_2$ photocathodes. *Chem. Mater.* **2017**, *29*, 4952–4962. [CrossRef]
17. Liu, Q.-L.; Zhao, Z.-Y.; Zhao, R.-D.; Yi, J.-H. Fundamental properties of delafossite $CuFeO_2$ as photocatalyst for solar energy conversion. *J. Alloys Compd.* **2020**, *819*, 153032. [CrossRef]
18. Preethi, S.; Vivek, S.; Priya, R.; Balakumar, S.; Babu, K.S. Enhanced photocatalytic performance of $CuFeO_2$-ZnO heterostructures for methylene blue degradation under sunlight. *J. Mater. Sci. Mater. Electron.* **2021**, *32*, 22256–22269. [CrossRef]
19. Tu, L.-W.; Chang, K.-S. Hydrothermal fabrication and photocatalytic study of delafossite ($CuFeO_2$) thin films on fluorine-doped tin oxide substrate. *Mater. Chem. Phys.* **2021**, *267*, 124620. [CrossRef]
20. Barnabe, A.; Mugnier, E.; Presmanes, L.; Thailhades, P. Preparation of delafossite $CuFeO_2$ thin films by rf-sputtering on conventional glass substrate. *Mater. Lett.* **2006**, *60*, 3468–3470. [CrossRef]
21. Zhu, T.; Deng, Z.; Fang, X.; Dong, W.; Shao, J.; Tao, R.; Wang, S. Room temperature deposition of amorphous p-type $CuFeO_2$ and fabrication of $CuFeO_2$/n-Si heterojunction by RF sputtering method. *Bull. Mater. Sci.* **2016**, *39*, 883–887. [CrossRef]
22. Ziani, N.; Aubry, E.; Martin, N.; Hirsinger, L.; Billard, A.; Briois, P.; Belkaid, M.S.; Yazdi, M.A.P. Influence of substrate temperature on delafossite $CuFeO_2$ films synthesized by reactive magnetron sputtering. *J. Alloys Compd.* **2021**, *876*, 160169. [CrossRef]
23. Li, S.Z.; Liu, J.; Wang, X.Z.; Yan, B.W.; Li, H.; Liu, J.-M. Epitaxial growth of delafossite $CuFeO_2$ thin films by pulse laser deposition. *Phys. B Condens. Mater.* **2012**, *407*, 2412–2415. [CrossRef]
24. Joshi, T.; Senty, T.R.; Trappen, R.; Zhou, J.; Chen, S.; Ferrari, P.; Borisov, P.; Song, X.; Holcomb, M.B.; Bristow, A.D.; et al. Structural and magnetic properties of epitaxial delafossite $CuFeO_2$ thin films grown by pulsed laser deposition. *J. Appl. Phys.* **2015**, *117*, 013908. [CrossRef]
25. Luo, S.; Fluri, A.; Zhang, S.; Liu, X.; Döbeli, M.; Harrington, G.F.; Tu, R.; Pergolesi, D.; Ishihara, T.; Lippert, T. Thickness-dependent microstructural properties of heteroepitaxial (00.1) $CuFeO_2$ thin films on (00.1) sapphire by pulsed laser deposition. *J. Appl. Phys.* **2020**, *127*, 065301. [CrossRef]
26. Zhang, L.; Li, P.; Huang, K.; Tang, Z.; Liu, G.; Li, Y. Chemical solution deposition and transport properties of epitaxial $CuFeO_2$ thin films. *Mater. Lett.* **2011**, *65*, 3289–3291. [CrossRef]
27. Gupta, R.K.; Cavas, M.; Al-Ghamdi, A.A.; Gafer, Z.H.; El-Tantawy, F.; Yakuphanoglu, F. Electrical and photoresponse properties of Al/p-$CuFeO_2$/p-Si/Al MTCOS photodiode. *Sol. Energy* **2013**, *92*, 1–6. [CrossRef]
28. Ito, M.; Izawa, C.; Watanabe, T. Direct fabrication of a $CuFeO_2$/Fe photocathode for solar hydrogen production by hydrothermal method. *Chem. Lett.* **2017**, *46*, 814–816. [CrossRef]

29. Yuan, J.; Yang, L.; Hao, C. Lithium-doped $CuFeO_2$ thin film electrodes for photoelectrochemical reduction of carbon dioxide to methanol. *J. Electrochem. Soc.* **2019**, *166*, H718. [CrossRef]
30. Aqaei, F.; Zare, M.; Shafiekhani, A. Role of plasmonic Au nanoparticles embedded in the diamond-like carbon overlayer in the performance of $CuFeO_2$ solar photocathodes. *J. Solid State Electrochem.* **2021**, *25*, 1139–1150. [CrossRef]
31. Yin, G.; Liu, C.; Shi, T.; Ji, D.; Yao, Y.; Chen, Z. Porous $BiVO_4$ coupled with $CuFeO_2$ and NiFe layered double hydroxide as highly-efficient photoanode toward boosted photoelectrochemical water oxidation. *J. Photochem. Photobiol. A Chem.* **2022**, *426*, 113742. [CrossRef]
32. Alkhayatt, A.H.O.; Thahab, S.M.; Zgair, I.A. Structure, surface morphology and optical properties of post-annealed delafossite $CuFeO_2$ thin films. *Optik* **2016**, *127*, 3745–3749. [CrossRef]
33. Garcia-Torregrosa, I.; Geertzema, Y.G.; Ismail, A.S.M.; Lee, T.-L.; de Grioot, F.M.F.; Weckhyusen, B.M. Facile two-step synthesis of delafossite $CuFeO_2$ photocathodes by ultrasonic spray pyrolysis and hybrid microwave annealing. *ChemPhotoChem* **2019**, *3*, 1238–1245. [CrossRef]
34. Dharmadasa, I.M.; Haigh, J. Strengths and advantages of electrodeposition as a semiconductor growth technique for applications in microelectronic devices. *J. Electrochem. Soc.* **2006**, *153*, G47–G52. [CrossRef]
35. Xiao, F.; Hangarter, C.; Yoo, B.; Rheem, Y.; Lee, K.-H.; Myung, N.V. Recent progress in electrodeposition of thermoelectric thin films and nanostructures. *Electrochim. Acta* **2008**, *53*, 8103–8117. [CrossRef]
36. Riveros, G.; Garin, C.; Ramirez, D.; Dalchiele, E.A.; Marotti, R.E.; Pereyra, C.J.; Spera, E.; Gomez, H.; Grez, P.; Martin, F.; et al. Delafossite $CuFeO_2$ thin films electrochemically grown from a DMSO based solution. *Electrochim. Acta* **2015**, *164*, 297–306. [CrossRef]
37. Kang, U.; Park, H. A facile synthesis of $CuFeO_2$ and CuO composite photocatalyst films for the production of liquid formate from CO_2 and water over a month. *J. Mater. Chem. A* **2017**, *5*, 2123–2131. [CrossRef]
38. Yan, Z.; Liu, H.; Hao, Z.; Yu, M.; Chen, X.; Chen, J. Electrodeposition of (hydro)oxides for an oxygen evolution electrode. *Chem. Sci.* **2020**, *11*, 10614–10625. [CrossRef] [PubMed]
39. Oh, Y.; Yang, W.; Tan, J.; Lee, H.; Park, J.; Moon, J. Photoelectrodes based on 2D opals assembled from Cu-delafossite double-shelled microspheres for an enhanced photoelectrochemical response. *Nanoscale* **2018**, *10*, 3720–3729. [CrossRef]
40. Oh, Y.; Yang, W.; Tan, J.; Lee, H.; Park, J.; Moon, J. Boosting visible light harvesting in p-type ternary oxides for solar-to-hydrogen conversion using inverse opal structure. *Adv. Funct. Mater.* **2019**, *29*, 1900194. [CrossRef]
41. Wang, M.; Liu, C.; Shi, H.; Long, T.; Zhang, C.; Liu, B. Facile synthesis of chitosan-derived maillard reaction productions coated $CuFeO_2$ with abundant oxygen vacancies for higher Fenton-like catalytic performance. *Chemosphere* **2021**, *283*, 131191. [CrossRef]
42. Choudhary, S.; Sarma, J.V.N.; Pande, S.; Girad, S.A.; Turban, P.; Lepine, B.; Gangopadhyay, S. Oxidation mechanism of thin Cu films: A gateway towards the formation of single oxide phase. *AIP Adv.* **2018**, *8*, 055114. [CrossRef]
43. Luo, J.; Steier, L.; Son, M.-K.; Schreier, M.; Mayer, M.T.; Grätzel, M. Cu_2O nanowire photocathodes for efficient and durable solar water splitting. *Nano Lett.* **2016**, *16*, 1848–1857. [CrossRef] [PubMed]
44. Mamyrbekova, A.K.; Mamitova, A.D.; Turebekova, G.; Gul, K.; Mamyrbekova, A.K. Kinetics and mechanism of cathodic processes at electrolysis of $Cu(NO_3)_2 \cdot 3H_2O$ solution in dimethyl sulfoxide. *Asian J. Chem.* **2016**, *28*, 525–528. [CrossRef]
45. De, M.; Gupta, S.P.S. Lattice imperfection studies in polycrystalline materials by x-ray diffraction line-profile analysis. *Pramana* **1984**, *23*, 721–744. [CrossRef]
46. Chen, H.-Y.; Fu, J.-R. Delafossite-$CuFeO_2$ thin films prepared by atmospheric pressure plasma annealing. *Mater. Lett.* **2014**, *120*, 47–49. [CrossRef]
47. Xiong, D.; Qi, Y.; Li, X.; Tao, H.; Chen, W.; Zhao, X. Hydrothermal synthesis of delafossite $CuFeO_2$ crystals at 100 °C. *RSC Adv.* **2015**, *5*, 49280–49286. [CrossRef]
48. Li, Z.; Wu, W.; Wang, M.; Wang, Y.; Ma, X.; Luo, L.; Chen, Y.; Kan, K.; Pan, Y.; Li, H.; et al. Ambient-pressure hydrogenation of CO_2 into long-chain olefins. *Nat. Commun.* **2022**, *13*, 2396. [CrossRef] [PubMed]
49. Son, M.-K.; Seo, H.; Watanabe, M.; Shiratani, M.; Ishihara, T. Characteristics of crystalline sputtered $LaFeO_3$ thin films as photoelectrochemical water splitting photocathodes. *Nanoscale* **2020**, *12*, 9653–9660. [CrossRef] [PubMed]

coatings MDPI

Article

Electrochemically Obtained Poly(3,4-ethylenedioxythiophene) Layers for Electroanalytical Determination of Lipoic Acid

Vasilena Karabozhikova and Vessela Tsakova *

Institute of Physical Chemistry, Bulgarian Academy of Sciences, Acad. G. Bonchev Str., Bl. 11, 1113 Sofia, Bulgaria
* Correspondence: tsakova@ipc.bas.bg

Abstract: Poly(3,4-ethylenedioxythiophene) (PEDOT) is a conducting polymer with intrinsic redox activity often used to facilitate electrooxidation reactions. PEDOT coatings with different thicknesses are obtained via electrochemical polymerization in the presence of either polysterensulfonate (PSS) or dodecylsulfate (SDS) anions. The electrooxidation of alfa lipoic acid (ALA) is studied depending on the thickness of the polymer coatings and the counterions used for their synthesis. The kinetics of ALA oxidation is found to differ for thin and thick PEDOT coatings with diffusion limitations observed for thin layers. For thick coatings, the rate-determining step varies from adsorption to diffusion depending on the ALA concentration. The type of counterion affects both the ALA oxidation peak currents and the peak potential. SDS-doped PEDOT coatings show a shift in the oxidation peak to positive potentials and higher ALA oxidation currents. The effect is commented in terms of a larger electroactive surface area and possible specific hydrophobic polymer/analyte interactions. For thin PEDOT coatings, the concentration dependence of the voltammetric peaks is linear in a wide concentration range (40 to 1000 μM), whereas the use of differential pulse voltammetry results in a linear response in a lower concentration range (8–200 μM) suitable for practical applications.

Keywords: conducting polymers; PEDOT; counterions; electroanalysis

Citation: Karabozhikova, V.; Tsakova, V. Electrochemically Obtained Poly(3,4-ethylenedioxythiophene) Layers for Electroanalytical Determination of Lipoic Acid. *Coatings* **2023**, *13*, 2014. https://doi.org/10.3390/coatings13122014

Academic Editors: Matjaž Finšgar and Peng Yu

Received: 24 October 2023
Revised: 22 November 2023
Accepted: 27 November 2023
Published: 28 November 2023

1. Introduction

Conducting polymers [1] are widely studied in view of various applications, including flexible electronics [2], biomedical applications [3–6], corrosion protection [7,8], supercapacitors [9], etc. The great variety of applications is due to the possibility to involve different aspects of the intrinsic characteristics of these materials, i.e., to use them as semiconductors, to exploit their ability for reversible redox switching or to use them in their oxidized, high-conducting state as coatings that provide a specific chemical environment.

Poly(3,4-ethylenedioxythiophene (PEDOT) is one of the most intensively investigated conducting polymer materials [10,11], providing good electrochemical stability and the possibility of synthesis in aqueous solutions. The polymerization of EDOT, either chemical or electrochemical, proceeds via the formation of oligomers and the further growth of partially oxidized polymeric chains that are positively charged. These positive charges become compensated for by counterions available in the synthesis solutions. Numerous investigations have demonstrated that the counterions affect various characteristics of PEDOT such as morphology and surface structure, spectral characteristics, type of ionic fluxes upon redox transition, and the rigidity of the coatings as well as their ability for swelling due to solvent uptake (see, e.g., [1,12–17]). Electrochemical polymerization is one of the easy ways to obtain PEDOT coatings on solid electrodes in the presence of different organic, inorganic or polymeric anions, providing, at the same time, the opportunity for the control of the polymerization charge (and thus the thickness of the polymer layer). The aim of the present study is to investigate the possibilities to involve PEDOT in the electroanalytical determination of alpha-lipoic acid (ALA) depending on the thickness of the polymer coatings and the counterions used in the course of its synthesis. Polysterenesulfonate (PSS) and dodecylsulfate (SDS) were chosen as counterions in the course of PEDOT electrochemical synthesis as these two anions provide polymer coatings with markedly different characteristics [13,14,16].

ALA, also known as thioctic acid, was chosen as a specific analyte that was neutral and underwent irreversible electrooxidation, also resulting in a neutral product (Scheme 1). Thus no coulomb interactions are expected between the analyte or product species and the polymer coating.

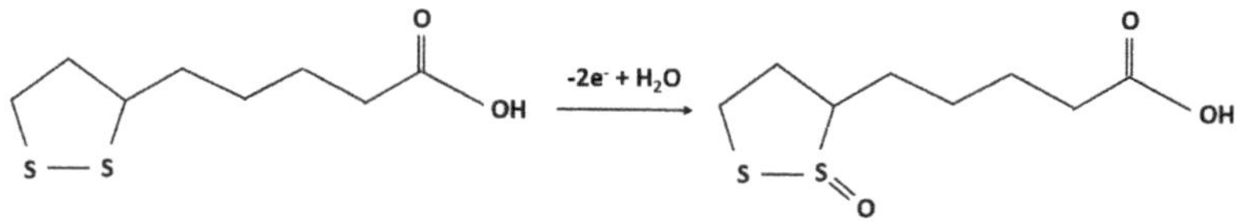

Scheme 1. Oxidation of ALA.

ALA is a natural cofactor and plays an important role in various multi-enzyme complexes. It is considered a universal anti-oxidant with anti-inflammatory properties and anti-cancer effects, and is used in therapies treating liver damage, heavy metal poisoning, diabetes, glaucoma, radiation damage, neurodegenerative diseases, etc. ([18,19] and the literature cited therein). Electroanalytical studies for ALA determination have already been carried out on various conventional, uncoated electrodes, such as glassy carbon (GCE) [18–20], pyrolitic graphite [21], boron-doped diamond [22], multiwall carbon nanotubes (MWCNT) [23], a carbon-fiber electrode [20], flour-doped tin oxide [24] and Pt [25]. Recently, more sophisticated electrode materials such as carboxilated MWCNT/polyindole/Ti_2O_3/glassy carbon [26] and SnO_2 nanoparticles- cetyltriphenylphosphonium bromide/GCE [27] were investigated for the electrochemical detection of ALA. Table 1 summarizes the data for the electroanalytical parameters resulting from these studies. The presented data show that the concentration range for the linear electroanalytical response is typically limited to about 200 μM L^{-1}, which corresponds to the practically relevant concentration range for ALA determination in foods and blood plasma specimens [28–30]. In two cases [23,27], a second linear range is reported, covering concentrations of up to 780 μM L^{-1}. The largest concentration interval of the linear response so was found on a Pt electrode [25]. The investigation of suitable alternative electrode materials could eventually open up the possibility to work in a wide concentration range with a single linear calibration curve by avoiding the use of noble metals and complex synthetic procedures necessary for electrode modifications. Conducting polymer coatings has so far not been involved in electroanalytical studies concerning ALA, although these coatings and specifically PEDOT provide suitable possibilities for the electrochemical sensing of different organic substances [31–33].

Table 1. Data for the linear range of the electroanalytical response and LOD obtained from different electrode materials.

Working Electrode	Electrolyte	Method	Linear Range [μM L^{-1}]	LOD [μM L^{-1}]	Reference
GCE	H_2SO_4	LSV DPV	1.51–173 2.5–75	5.75 1.8	[18]
GCE	PBS	DPV	1–150	1.8	[19]
MWCNTs/GCE	H_2SO_4	LSV	26–180 * 210–780	19	[23]
Co-phtalocynine/pyrolytic graphite	PBS	DPV CA CV	0.499–19.6 1.9–25 7.3–260	0.0034 0.098 0.25	[21]
B-doped diamond	BRB, pH3	DPV CA	0.3–105 0.3–60	0.088 0.06	[22]
F-doped SnO_2	PBS	SWV	5–200	3.7	[24]

Table 1. *Cont.*

Working Electrode	Electrolyte	Method	Linear Range [μM L^{-1}]	LOD [μM L^{-1}]	Reference
Pt	BRB, pH 4.5	DPV	10–800	13.15	[25]
MWCNT/-polyindole/Ti_2O_3	PBS	DPV	0.39–110	0.0012	[26]
SnO_2 NPs-CTPPB/GCE	BRB, pH 4.5	DPV	0.5–50 * 50–400	0.13 0.43	[27]

* two linear ranges. BRB—Briton–Robinson buffer; CA—chronoamperometry; CTPPB—cetyltriphenylphosphonium bromide; LSV—linear sweep voltammetry; PBS—phosphate-buffered solution; SWV—square wave voltammetry.

In the present work, PSS- and SDS-doped PEDOT layers with two different thicknesses (several tens of nanometers and several hundreds of nanometers as described below) are obtained via electrochemical polymerization and further studied as electrode coatings for the electrooxidaton of ALA. The aim of this investigation is to reveal the role of the polymer layer's thickness and doping ions for the electrooxidation reaction and to explore the possibilities for the use of PEDOT-coated electrodes for the electroanalytical detection of ALA.

2. Materials and Methods

2.1. Experimental Set-Up

The electrochemical measurements were performed with an Autolab potentiostat/galvanostat (Metrohm-Autolab, Utrect, The Netherlands) in a three-electrode configuration. The working electrode was a glassy carbon rod embedded in resin with a front surface area of 0.070 cm^2. Before each EDOT polymerization experiment, the glassy carbon electrode (GCE) was polished using P2000 and P4000 grinding papers (Buehler, Lake Bluff, IL, USA), followed by 0.25 mm alumina suspension (Buehler Topol 3). A platinum plate was used as a counter electrode. The reference electrode was a mercury/mercury sulfate/0.5 mol L^{-1} K_2SO_4 electrode (MSE) with E_{MSE} = 0.66 V vs. SHE. All potentials in the text are referred to as the MSE. The electrolyte solutions were de-aerated using argon before use.

2.2. Synthesis of PEDOT Coatings

The PEDOT layers were obtained via potentiostatic polymerization at E = 0.38 V in aqueous solutions of 10 mmol L^{-1} of EDOT and 34 mmol L^{-1} of either sodium PSS (Mw = 70,000; 30% aqueous solution) or SDS (in the absence of additional inorganic salt in the solution). The PEDOT coatings obtained in the presence of either type of counterions will be further denoted as PEDOT-PSS and PEDOT-SDS. The charge, Q_{poly}, measured in the course of polymerization was used to control the thickness of the PEDOT coatings. Q_{poly} was fixed at two values, 1 mC (i.e., 14 mC cm^{-2}) and 8 mC (i.e., 112 mC cm^{-2}), denoted further as "thin" and "thick" PEDOT coatings. A direct recalculation of the polymerization charge in polymer layer thickness is usually not feasible because of the known influence of the counterions used in the course of polymerization on the structure, and the morphology and extent of swelling of the polymer coatings. A rough estimate for the SDS-doped coatings can be made based on the reference value obtained via a direct AFM measurement of a PEDOT (ClO_4^--doped) layer resulting in a 600 nm thickness for a polymerization charge of 114 mC cm^{-2} [13]. Thus, the thickness of the "thick" PEDOT-SDS coatings should be of that order of magnitude whereas the "thick" PSS-doped coatings are expected to have a thickness that is larger by a factor of 1.35 [14] and to amount to about 800 nm.

2.3. Electro-Oxidation of ALA

Ther electro-oxidation of ALA was studied in acetate buffer solution at pH 5 via both cyclic voltammetry (CV) and differential pulse voltammetry (DPV). A series of measurements was first performed in phosphate-buffered solution (pH 7) but the ALA oxidation peak was overlapping with the intrinsic overoxidation peak of PEDOT, and ALA oxidation currents, if detectable, were smaller than those in the acetate buffer at pH 5.

Before electroanalytical measurements were performed, the PEDOT-coated electrodes were conditioned in the buffer solution via several voltammetric scans until a stable voltammetric curve was obtained. Typically, 10 scans were needed to obtain a stable voltammetric response. The PEDOT-coated electrodes were stored in acetate buffer.

The parameters of the DPV procedure were a step potential of 5 mV, modulation amplitude of 20 mV, modulation time of 0.05 s, and interval time of 0.5 s. Capsules in an amount of 300 mg (Thioctic acid, Adipharm, Bulgaria) were used for real sample measurement. The content of the capsule was first dissolved in ethanol and further diluted in acetate buffer. The standard addition method was used for the determination of ALA in an electrochemical cell with 90 mL of acetate buffer. The limit of detection (LOD) was calculated by using the 3 s/m criterion, where s is the standard deviation of the linear regression intercept and m is the slope of the analytical calibration curve.

3. Results and Discussion

3.1. Electrochemical Synthesis of the PEDOT Coatings

Figure 1 shows several polymerization curves of PEDOT obtained in the presence of both types of counterions. In general, polymerization in the presence of SDS is faster but also shows a larger data discrepancy. Bearing in mind that the GCE is mechanically polished before polymerization, this effect should relate to the role of the initial GCE surface state for the PEDOT-SDS layer's formation. Polymerization in the presence of PSS is reproducible and not that susceptible to the initial surface state of the glassy carbon.

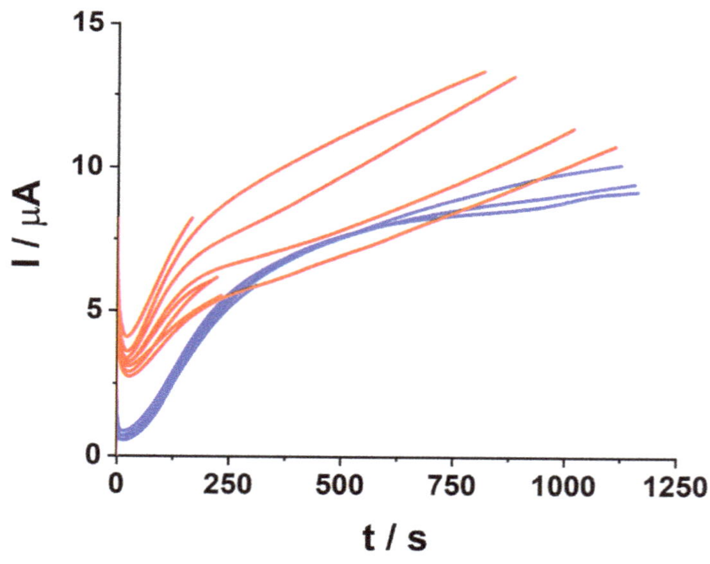

Figure 1. Polymerization curves obtained at E = 0.38 V vs. MSE in aqueous solution of 10 mmol L^{-1} of EDOT and 34 mmol L^{-1} of either PSS (blue lines) or SDS (red lines). The current traces registered for short times (200–275 s) correspond to Q_{poly} = 1 mC.

The voltammetric behavior of both thin and thick layers of SDS- and PSS-doped PEDOT is shown in Figure 2. Twenty voltammetric scans are necessary to establish steady-state behavior after polymerization. For both polymerization charges (Q_{poly} = 1 mC and Q_{poly} = 8 mC), PEDOT-PSS shows smaller pseudo-capacitive currents than PEDOT-SDS does, which implies a smaller electroactive surface in the PSS case. SEM pictures (Figure 3) demonstrate the difference in the surface morphology of the two types of PEDOT layers.

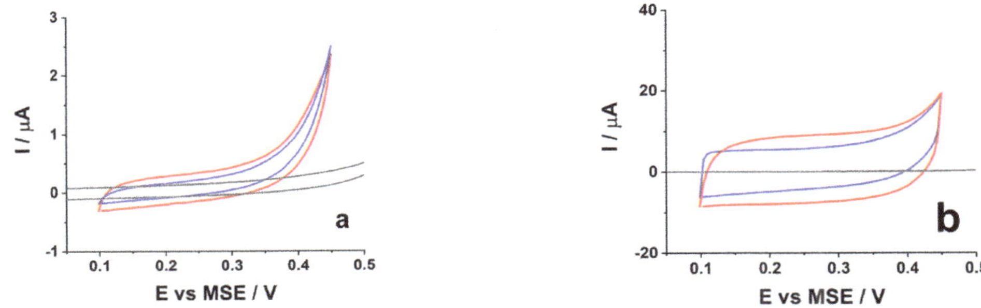

Figure 2. Cyclic voltammetry measured in acetate buffer solution at 50 mV s^{-1} for (**a**) thin (Q_{poly} = 1 mC) and (**b**) thick (Q_{poly} = 8 mC) PEDOT-PSS (blue lines) and PEDOT-SDS (red lines) coatings. The grey lines denote the trace of the un-coated glassy carbon electrode.

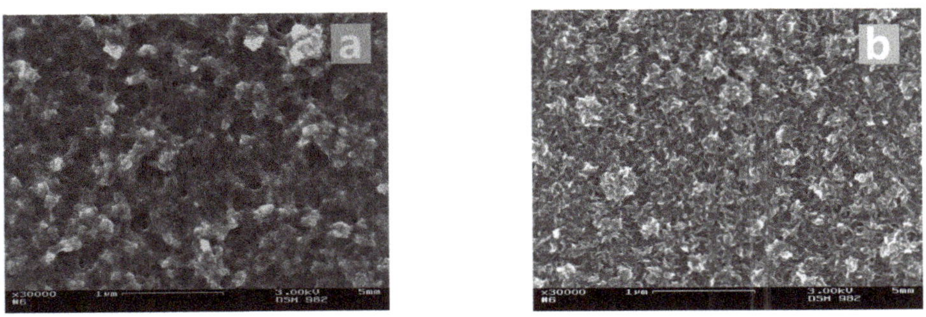

Figure 3. SEM pictures of (**a**) PEDOT-PSS and (**b**) PEDOT-SDS coatings. Reprinted with permission from Electrochimica Acta, Vol. 122, p.21, V. Lyutov, I. Efimov, A. Bund, V. Tsakova Electrochemical polymerization of 3,4-ethylenedioxythiophene in the presence of dodecylsulfate and polysulfonic anions—an acoustic impedance study [13], Copyright Elsevier, 2014.

In the next series of experiments, the scan rate dependence of the voltammetric response was measured in acetate buffer solution for all PEDOT coatings. Figure 4 presents results obtained for the thick PEDOT-PSS and PEDOT-SDS coatings. The double logarithmic plot (Figure 4c) shows a linearization of the data for the current dependent on the scan rate with the values of the straight line slopes in both cases being close to 1. The latter is also valid for the thin PEDOT layers and indicates usual capacitive behavior without the interference of limiting ionic diffusion inside the polymer coatings.

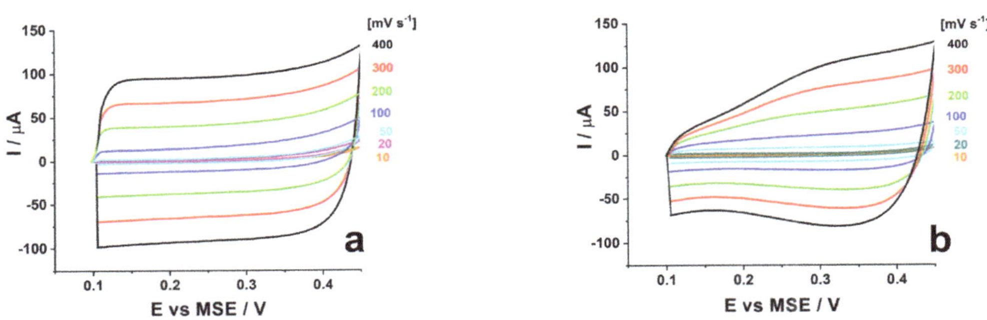

Figure 4. *Cont.*

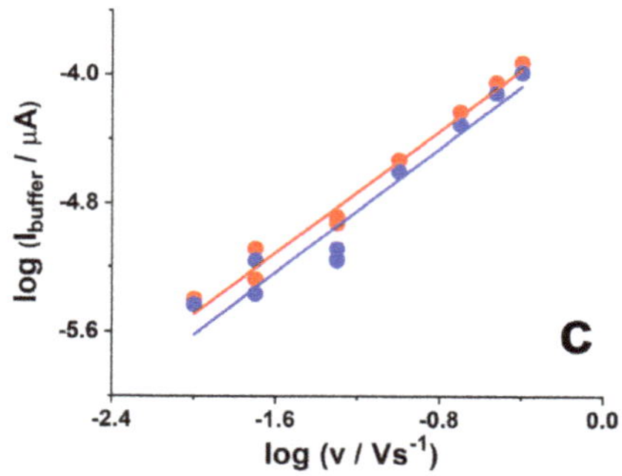

Figure 4. Scan rate dependence of the intrinsic pseudocapacitive currents of thick (Q_{poly} = 8 mC) (**a**) PEDOT-PSS and (**b**) PEDOT-SDS coatings measured in acetate buffer solution; (**c**) double logarithmic plot of the data obtained for PEDOT-PSS (blue symbols) and PEDOT-SDS (red symbols).

3.2. Voltammetric Measurements in the Presence of ALA

A comparative study of the electrooxidation of ALA on PSS- and SDS-doped PEDOT with the two different thicknesses was first carried out by cyclic voltammetry. Figure 5 shows sets of voltammetric curves measured at different ALA concentrations ranging from 40 µmol L^{-1} to 1 mmol L^{-1}. In all cases the ALA oxidation peak is well detectable but there are marked differences depending on the doping ions and the thickness of the PEDOT layers. For PEDOT-PSS the ALA oxidation peak is found at less positive potentials whereas for the SDS-doped ones it shifts by about 60 mV in positive direction which implies an inhibited electron transfer in the latter case. Nevertheless, the oxidation peak currents are higher in the SDS case due very probably to the larger electroactive surface area and possible specific analyte/coating interactions (see below).

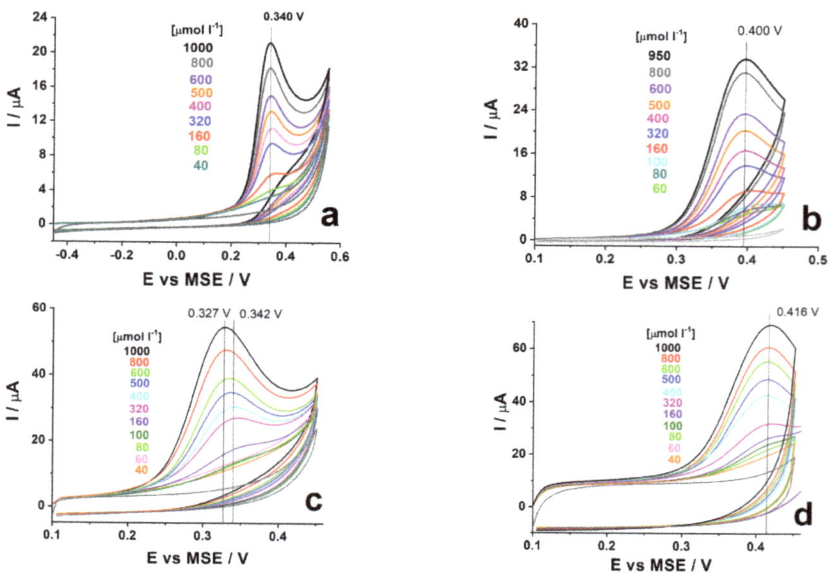

Figure 5. Cyclic voltammetry in the presence of different amounts of ALA measured from (**a**) thin (Q_{poly} = 1 mC) PEDOT-PSS, (**b**) thin (Q_{poly} = 1 mC) PEDOT-SDS, (**c**) thick (Q_{poly} = 8 mC) PEDOT-PSS, and (**d**) thick (Q_{poly} = 8 mC) PEDOT-SDS coatings. Scan rate: 50 mV s^{-1}. Grey lines denote measurements in buffer solution alone.

The comparison between thick and thin PEDOT layers at a constant ALA concentration shows, for both types of counterions, higher peak current values for the thicker coatings that should be related to a higher electroactive surface area of coatings with Q_{poly} = 8 mC. It is well established that with the thickening of the conducting polymer layers, there is a change in surface morphology, resulting in a rougher and/or more open and porous structure. Both effects would lead to an increase in the effective electroactive surface area of the coatings. It is worth mentioning that for the PEDOT-SDS and the thin PEDOT-PSS layers, the peak potential remains fixed for all concentration measured (Figure 5a,b,d). For the thick PEDOT-PSS layer (Figure 5c), however, a shift in the peak position by about 15 mV in the negative direction is observed. The latter seems to result from the gradual suppression of the PEDOT-PSS intrinsic oxidation currents found at E > 0.35 V.

A voltammetric scan rate study was carried out also in the presence of ALA in the acetate buffer solution. The double-logarithmic plot of the data for the oxidation peak currents vs. scan rate (Figure 6) obtained for the thin PEDOT coatings shows linearization with slopes of the straight lines amounting to 0.57 and 0.65 for PEDOT-PSS and PEDOT-SDS, respectively (Figure 6a). The latter values are close to the expected value (0.5) for a diffusion-limited process. The measurements with the thick PEDOT coatings (Figure 6b, full symbols) show a different result. At the 100 µmol L^{-1} ALA concentration, the slope of the log(I) vs. log(v) dependence amounts to 1, whereas by increasing the concentration of ALA up to 400 µmol L^{-1} (Figure 6b, hollow squares), the slope decreases to a value (0.7) closer to the one expected for diffusion. This peculiarity means that for thick PEDOT layers, a linear calibration curve, valid for both low and high concentration ranges of ALA, should not be expected.

Finally, the concentration dependences of the peak currents for the four types of PEDOT coatings are shown in Figure 7. In order to eliminate the contribution of the intrinsic PEDOT-related currents, the data for the ALA oxidation peaks are adjusted by subtracting the corresponding values of the buffer currents of the respective PEDOT coatings. The resulting concentration dependences show higher sensitivities for the thick PEDOT coatings. Larger values of sensitivity are observed for PEDOT-SDS for both thicknesses (see Table 2). The values of LOD obtained by means of the different PEDOT specimens using cyclic voltammetry range from 34 to 50 µmol L^{-1}.

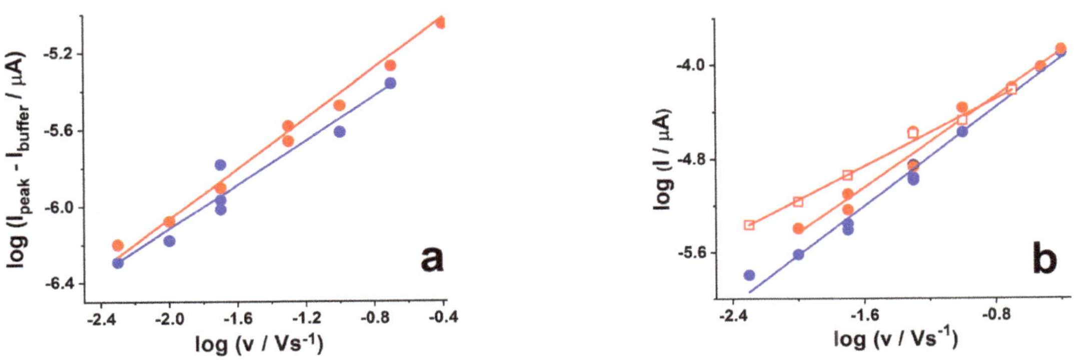

Figure 6. Double logarithmic plot of data obtained from the scan rate dependence measured in acetate buffer in the presence of 100 µmol of L^{-1} ALA for PEDOT-PSS (blue symbols) and PEDOT-SDS (red symbols): (**a**) thin (Q_{poly} = 1 mC) and (**b**) thick (Q_{poly} = 8 mC) PEDOT coatings. The red squares in the plot (**b**) are obtained with 400 µmol L^{-1} of ALA in the solution.

Table 2. Data for the sensitivity for ALA oxidation and LOD obtained from different PEDOT coatings.

Method	Specimen	Concentration Range [μmol L^{-1}]	Sensitivity [μA $μmol^{-1}$ cm^{-2}]	LOD [μmol L^{-1}]
CV	PSS, 1 mC	40–1000	0.233	50
CV	SDS, 1 mC	60–950	0.413	36
CV	PSS, 8 mC	100–1000	0.550	34
CV	SDS, 8 mC	100–750	0.876	40
DPV	PSS, 1 mC	8–200	0.146	6
DPV	SDS, 1 mC,	10–100	0.163	7

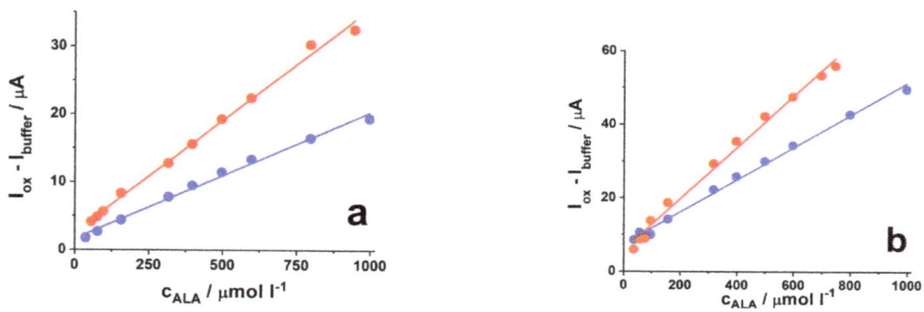

Figure 7. ALA peak oxidation currents, after the extraction of the corresponding buffer current values, for (**a**) thin (Q_{poly} = 1 mC) and (**b**) thick (Q_{poly} = 8 mC) PEDOT coatings, PEDOT-PSS (blue symbols), and PEDOT-SDS (red symbols).

3.3. DPV Measurements in the Presence of ALA

It is well known that the DPV method provides the possibility of increasing the sensitivity of the electroanalytical measurements, and thus allows for electroanalytical detection in a lower concentration range. Figure 8 shows a series of DPV measurements at different ALA concentrations carried out with thin PEDOT-PSS and PEDOT-SDS coatings. The DPV signal is sensitive to ALA concentrations as low as about 10 μmol L^{-1} for both types of electrodes and close values of the sensitivities are found for both doping anions (Figure 9 and Table 2). The regression lines for both types of PEDOT are $y = -2.27 \times 10^{-8} + 0.0117x$ (R-square = 0.9959) for PSS and $y = -6.42 \times 10^{-8} + 0.0131x$ (R-square = 0.9890) for SDS. The DPV method could not be applied for the thick PEDOT coatings, very possibly due to the fact that for these coatings, the limiting factor controlling the ALA oxidation reaction in the low concentration range is not diffusion.

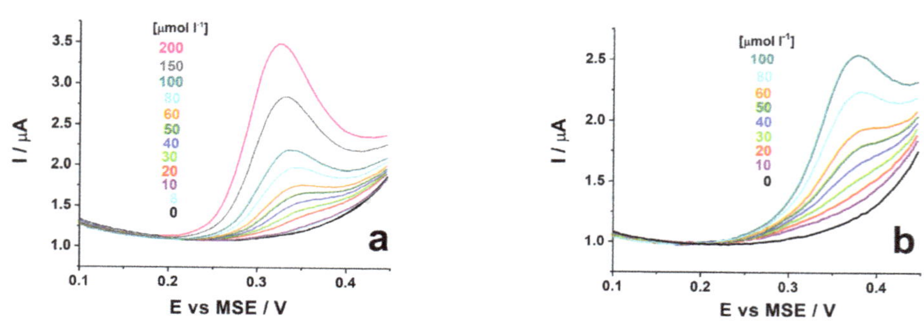

Figure 8. DPV measurements in acetate buffer solution at different concentrations of ALA for thin (Q_{poly} = 1 mC) (**a**) PEDOT-PSS and (**b**) PEDOT-SDS coatings.

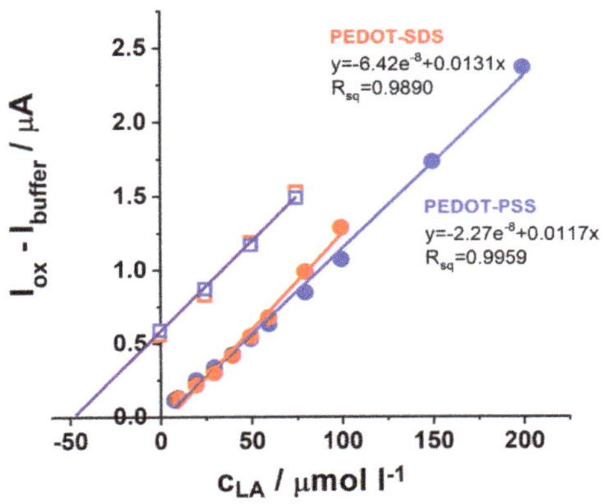

Figure 9. Calibration curves obtained from the DPV concentration dependences from thin (Q_{poly} = 1 mC) PEDOT-PSS (blue points) and PEDOT-SDS (red points) layers. The blue and red squares show the results from the determination of ALA in a real specimen via the standard addition method using the respective polymer coatings.

The thin PEDOT coatings were further used for the determination of ALA in real samples. A commercially available food supplement was used for this aim. Via dilution (see Section 2.3) an amount corresponding to a concentration of 50 μmol L^{-1} was introduced in the buffer solution. Data obtained via the standard addition method are presented in Figure 10. The data for the current obtained in these measurements show a linear dependence on the added concentration with straight lines parallel to the initial calibration curves (Figure 9 open symbols). The intercept from the X-axis results in ALA concentration values corresponding to 49 μmol L^{-1}, i.e., to 98% of the expected value.

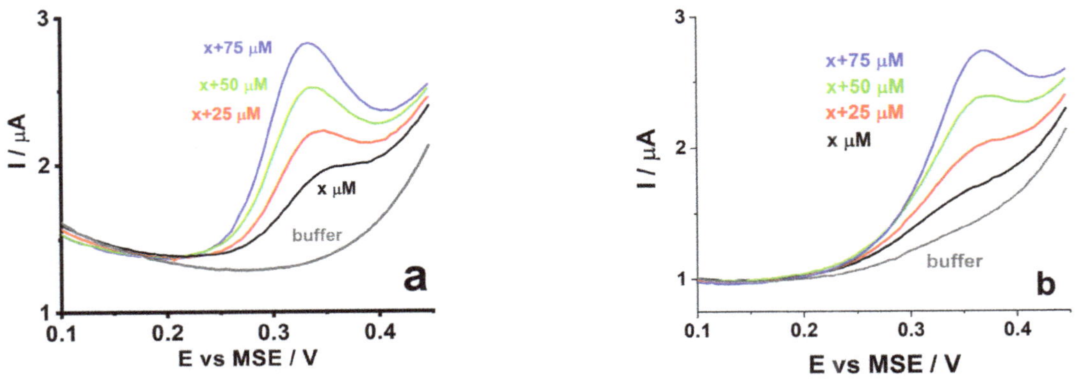

Figure 10. DPV measurements in acetate buffer solution obtained via the standard addition method from thin (Q_{poly} = 1 mC) (**a**) PEDOT-PSS and (**b**) PEDOT-SDS coatings.

4. Conclusions

The present investigation demonstrates the possibility to involve PEDOT in electroanalytical determination of ALA by focusing on the characteristics, thickness and type of counterions that influence the electrooxidation reaction. It is found that by increasing the thickness of PEDOT, irrespective of the counterions, the oxidation currents of ALA increase by more than twice, most probably due to the real surface area of the polymer coatings. An important point in the present investigation is the established difference in the kinetics of the ALA oxidation reaction depending on the thickness of the PEDOT coatings and the concentration of ALA. In fact, voltammetric or DPV peak currents should depend linearly on the analyte concentration, provided that diffusion is limiting the electrochemical reaction in the whole studied concentration interval. This is obviously the case for the thin PEDOT coatings. For the thick PEDOT coatings, however, the enlarged surface area, which is beneficial for increasing the sensitivity of the electroanalytical signal, results simultaneously in the opportunity to experience kinetic limitations, seemingly due to the adsorption of analyte or product species. If adsorption alone is limiting the electrochemical reaction, non-linear calibration curves reaching saturation are observed. In the present case, adsorption is expected to limit the ALA oxidation reaction on thick PEDOT layers and at a low analyte concentration. Nevertheless, at concentrations above 100 μmol L^{-1}, diffusion starts to be operative as far the polymer electroactive surface seems to be saturated with respect to dynamic adsorption/desorption but not blocked with respect to electron transfer and further oxidation events. Thus, the rate-determining steps in the low and high concentration ranges differ and the thick PEDOT coatings could not be properly used at very low ALA concentrations despite the higher sensitivity of the electroanalytical signal. For thin PEDOT layers, however, a large concentration range (40 to 1000 μM L^{-1}) of the linear voltammetric response is established.

Concerning the type of the counter ions in PEDOT, a marked effect is observed in the electrooxidation reaction. In the SDS case, a shift in the ALA oxidation peaks to more positive potentials is established together with larger oxidation currents. Apart from the difference in surface morphology and real electroactive surface area, specific hydrophobic/hydrophilic interactions of the analyte species with the two types of PEDOT coatings should be considered. PSS-doped PEDOT is a more hydrophilic material with significant swelling in aqueous solutions and is also in the origin of the larger thickness at a constant polymerization charge. On the other hand, SDS-doped PEDOT layers are hydrophobic and expected to involve small amounts of solvent molecules in their internal structure. ALA has both a hydrophobic and hydrophilic part but charge transfer should occur in the hydrophobic one. Thus, PEDOT-SDS coatings should provide a better chemical environment for the ALA oxidation reaction. At the same time, a shift in the potential positions of the ALA oxidation peaks depending on the type of counterions provides the opportunity for eventually avoiding the possible overlap of oxidation peaks due to interfering species.

Finally, compared to those obtained in previous electroanalytical studies conducted for ALA determination (Table 1), the present study's data for the sensitivity, LOD and width of the linear concentration range (Table 2) show the possibility to work in a very large concentration range. The DPV measurements performed with the thin PEDOT layers provide the possibility to work at concentrations even below 10 μmol L^{-1}, and to perform the electroanalytical determination of ALA in real specimens. Concentrations of several μmol L^{-1} to several tens of μmol L^{-1} are of practical importance for the determination of ALA in foods [28] and blood plasma [29,30], whereas pharmaceutical formulations and food additives have much larger concentrations [26,27]. Thus, the present study shows that the intrinsic characteristics of PEDOT such as thickness and type of counterions provide a useful instrument for achieving different electroanalytical performance and addressing various tasks concerning the electrochemical determination of ALA.

Author Contributions: Conceptualization, V.T.; methodology, V.T.; investigation, V.K.; analysis of data, V.K. and V.T.; resources, V.T.; writing—original draft preparation, V.T.; writing—review and editing, V.T.; visualization, V.K.; supervision, V.T.; funding acquisition, V.T. All authors have read and agreed to the published version of the manuscript.

Funding: This research was funded by the Bulgarian OP Science and Education for Smart Growth (2014–2020) within National Center of Mechatronics and Clean Technologies (Project BG05M2OP001-1.001-0008-C010) and Bulgarian Ministry of Education and Science within INFRAMAT (National Roadmap for Research Infrastructure).

Institutional Review Board Statement: Not applicable.

Informed Consent Statement: Not applicable.

Data Availability Statement: Data are contained within the article.

Acknowledgments: The technical support of A. Nakova (Institute of Physical Chemistry, Bulgarian Academy of Sciences) is gratefully acknowledged.

Conflicts of Interest: The authors declare no conflict of interest. The funders had no role in the design of the study; in the collection, analyses, or interpretation of data; in the writing of the manuscript; or in the decision to publish the results.

References

1. Inzelt, G. *Conducting Polymers: A New Era in Electrochemistry*; Springer: Berlin/Heidelberg, Germany, 2012.
2. Ouyang, J. Application of intrinsically conducting polymers in flexible electronics. *SmartMat* **2021**, *2*, 263. [CrossRef]
3. Bobacka, J.; Ivaska, A. Chemical sensors based on conducting polymers. In *Electropolymerization. Concepts, Materials, Applications*; Cosnier, S., Karyakin, A., Eds.; Wiley: Weinheim, Germany, 2010; pp. 173–188.
4. Tsakova, V. Conducting polymers in electroanalytical medical applications. In *Applications of Electrochemistry in Medicine, Modern Aspects of Electrochemistry*; Schlesinger, M., Ed.; Springer Science + Business Media: New York, NY, USA, 2013; Volume 56, pp. 283–342.
5. Tezakati, T.; Seifalian, A.; Tan, A.; Seifalian, A.M. Conductive Polymers: Opportunities and Challenges in Biomedical Applications. *Chem. Rev.* **2018**, *118*, 6766. [CrossRef] [PubMed]
6. Ramanavicius, S.; Ramanavicius, A. Conducting polymers in the Design of Biosensors and Biofuel Cells (Review). *Polymers* **2021**, *13*, 49. [CrossRef] [PubMed]
7. Xu, H.; Zhang, J. A Review on Conducting Polymers and Nanopolymer Composite Coatings for Steel Corrosion Protection. *Coatings* **2019**, *9*, 807. [CrossRef]
8. Umoren, S.A.; Solomon, M.M. Protective polymeric films for industrial substrates: A critical review on past and recent applications with conducting polymers and polymer composites/nanocomposites. *Progr. Mat. Sci.* **2019**, *104*, 38. [CrossRef]
9. Zhao, Z.; Xia, K.; Hou, Y.; Zhang, Q.; Yead, Z.; Lu, J. Designing flexible, smart and self-sustainable supercapacitors for portable/wearable electronics: From conductive polymers. *Chem. Soc. Rev.* **2021**, *50*, 12702–12743. [CrossRef] [PubMed]
10. Groenendaal, L.; Jonas, F.; Freitag, D.; Pielartzik, H.; Reynolds, J.R. Poly(3,4-ethylenedioxythiophene) and its derivatives: Past, present, and future. *Adv. Mater.* **2000**, *12*, 481–494. [CrossRef]
11. Elschner, A.; Kirchmeyer, S.; Lovenich, W.; Merker, U.; Reuter, K. *PEDOT: Principles and Applications of an Intrinsically Conductive Polymer*; CRC Press: Boca Raton, FL, USA, 2010.
12. Nasybulin, E.; Wei, S.; Kymissis, I.; Levon, K. Effect of solubilizing agent on properties of poly(3,4-ethylenedioxythiophene) (PEDOT) electrodeposited from aqueous solution. *Electrochim. Acta* **2012**, *78*, 638. [CrossRef]
13. Lyutov, V.; Efimov, I.; Bund, A.; Tsakova, V. Electrochemical polymerization of 3,4-ethylenedioxythiophene in the presence of dodecylsulfate and polysulfonic anions—An acoustic impedance study. *Electrochim. Acta* **2014**, *122*, 21. [CrossRef]
14. Lyutov, V.; Gruia, V.; Efimov, I.; Bund, A.; Tsakova, V. An acoustic impedance study of PEDOT layers obtained in aqueous solution. *Electrochim. Acta* **2016**, *190*, 285. [CrossRef]
15. Gribkova, O.L.; Iakobson, O.D.; Nekrasov, A.A.; Cabanova, V.A.; Tverskoy, V.A.; Vannikov, A.V. The influence of polyacid nature on poly(3,4-ethylenedioxythiophene) electrosynthesis and its spectroelectrochemical properties. *J. Solid State Electrochem.* **2016**, *20*, 2991. [CrossRef]
16. Lyutov, V.; Kabanova, V.; Gribkova, O.; Nekrasov, A.; Tsakova, V. Electrochemically Obtained Polysulfonates Doped Poly(3,4-ethylenedioxythiophene) Films—Effects of the Dopant's Chain Flexibility and Molecular Weight Studied by Electrochemical, Microgravimetric and XPS Methods. *Polymers* **2021**, *13*, 2438. [CrossRef] [PubMed]
17. Gueyea, M.N.; Carella, A.; Faure-Vincent, J.; Demadrillec, R.; Simonato, J.-P. Progress in understanding structure and transport properties of PEDOT-based materials: A critical review. *Progr. Mater. Sci.* **2020**, *108*, 100616. [CrossRef]
18. Ziyatdinova, G.K.; Budnikov, G.K.; Pogoreltsev, V.I. Electrochemical determination of lipoic acid. *J. Analyt. Chem.* **2004**, *59*, 288. [CrossRef]
19. Corduneanu, O.; Garnett, M.; Oliveira Brett, A.M. Anodic Oxidation of a-Lipoic Acid at a Glassy Carbon Electrode and Its Determination in Dietary Supplements. *Analyt. Lett.* **2007**, *40*, 1763. [CrossRef]
20. Skorupa, A.; Michalkiewicz, S. Anodic Oxidation of α-Lipoic Acid on Carbon Electrodes in Acetic Acid—Acetonitrile Solutions, Int. *J. Electrochem. Sci.* **2019**, *14*, 5107. [CrossRef]
21. Ferreira, A.P.M.; dos Santos Pereira, L.N.; Santos da Silva, I.; Tanaka, S.M.C.N.; Tanaka, A.A.; Angnes, L. Determination of a-Lipoic acid on a Pyrolytic Graphite Electrode Modified with Cobalt Phthalocyanine. *Electroanalysis* **2014**, *26*, 2138. [CrossRef]
22. Stankovic, D.M.; Mehmeti, E.; Kalcher, K. Development of sensitive electroanalytical approach for the quantification of α-lipoic acid using boron doped diamond. *Analyt. Sci.* **2016**, *32*, 847. [CrossRef]
23. Ziyatdinova, G.K.; Grigoreva, L.V.; Budnikov, G.K. Electrochemical Determination of Unithiol and Lipoic Acid at Electrodes Modified with Carbon Nanotubes. *J. Analyt. Chem.* **2009**, *64*, 185. [CrossRef]

24. Miranda, M.P.; del Rio, R.; del Valle, M.A.; Faundez, M.; Armijo, F. Use of fluorine-doped tin oxide electrodes for lipoic acid determination in dietary supplements. *J. Electroanal. Chem.* **2012**, *668*, 1. [CrossRef]
25. Marin, M.; Lete, C.; Manolescu, B.N.; Lupu, S. Electrochemical determination of a-lipoic acid in human serum at platinum electrode. *J. Electroanal. Chem.* **2014**, *729*, 128. [CrossRef]
26. Sasikumara, R.; Ranganathan, P.; Chen, S.-M.; Rwei, S.-P. f-MWCNTs-PIN/Ti_2O_3 nanocomposite: Preparation, characterization and nanomolar detection of α-Lipoic acid in vegetables. *Sens. Actuators B* **2018**, *255*, 217. [CrossRef]
27. Ziyatdinova, G.; Antonova, T.; Vorobev, V.; Osin, Y.; Budnikov, H. Selective voltammetric determination of α-lipoic acid on the electrode modified with SnO_2 nanoparticles and cetyltriphenylphosphonium bromide. *Monatshefte Für Chem.-Hemical Mon.* **2019**, *150*, 401. [CrossRef]
28. Saha, M.; Ahammad, H.; Bhoumik, N.; Shakil, S.; Shavan, M.; Morshed, M.; Hossan, T.; Sarker, S.; Rahman, N.; Rahman, S.; et al. Extraction and estimation of alpha lipoic acid content in different food samples by reverse phase HPLC: Effect of heat treatment. *Int. J. Biosci.* **2018**, *13*, 473.
29. Amenta, F.; Traini, E.; Tomassoni, D.; Mignini, F. Pharmacokinetics of Different Formulations of Tioctic (Alpha-Lipoic) Acid in Healthy Volunteers. *Clin. Exp. Hypertens.* **2008**, *30*, 767. [CrossRef]
30. Shay, K.; Moreau, R.; Smith, E.; Smith, A.; Hagen, T. Alpha-lipoic acid as a dietary supplement: Molecular mechanisms and therapeutic potential. *Biochim. Biophys. Acta* **2009**, *1790*, 1149. [CrossRef] [PubMed]
31. Zanardi, C.; Terzi, F.; Seeber, R. Polythiophenes and polythiophene-based composites in amperometric sensing. *Analyt. Bioanalyt. Chem.* **2013**, *405*, 509. [CrossRef]
32. Tsakova, V.; Seeber, R. Conducting polymers in electrochemical sensing: Factors influencing the electroanalytical signal. *Analyt. Bioanalyt. Chem.* **2016**, *408*, 7231. [CrossRef]
33. Hui, Y.; Bian, C.; Xia, S.; Tong, J.; Wang, J. Synthesis and electrochemical sensing application of poly(3,4-ethylenedioxythiophene)-based materials: A review. *Analyt. Chim. Acta* **2018**, *1022*, 1. [CrossRef]

coatings

Article

Medium and High Phosphorous Ni-P Coatings Obtained via an Electroless Approach: Optimization of Solution Formulation and Characterization of Coatings

Virgilio Genova [1,*], Laura Paglia [1], Giovanni Pulci [1], Giulia Pedrizzetti [1], Alice Pranzetti [2], Marco Romanelli [2] and Francesco Marra [1]

[1] Department of Chemical Engineering, Materials, Environment, Sapienza University of Rome, INSTM Reference Laboratory for Engineering of Surface Treatments, Via Eudossiana 18, 00184 Rome, Italy; laura.paglia@uniroma1.it (L.P.); giovanni.pulci@uniroma1.it (G.P.); giulia.pedrizzetti@uniroma1.it (G.P.); francesco.marra@uniroma1.it (F.M.)
[2] Nuovo Pignone International, Baker Hughes, Via Felice Matteucci 2, 50127 Florence, Italy; alice.pranzetti@bakerhughes.com (A.P.); marco.romanelli@bakerhughes.com (M.R.)
* Correspondence: virgilio.genova@uniroma1.it; Tel.: +39-(0)644-585-314

Abstract: A new lead-free electroless Ni-P plating solution was developed for the deposition of coatings with medium phosphorus content (MP, 6–9 wt%), and its composition was optimized to obtain deposits with high phosphorus (HP, 10–14 wt%). Cleaning and activation treatments were studied in terms of effectiveness and influence on the deposition rate. The concentration of reagents (nickel salt, complexing agent, reducing agent and stabilizer) was studied, and their combined effect on P content and plating rate was investigated. The obtained coatings were analyzed by SEM and XRD and thermally treated at 400 °C and 600 °C to study microstructural evolution. Vickers hardness was measured on as-deposited and annealed coatings to relate hardness evolution to microstructural changes after thermal treatments. Optimal deposition conditions were determined, enabling the production of MP coatings (6.5 wt% P) with a plating rate of 40 µm/h and HP coatings (10.9 wt% P) with a plating rate of 25 µm/h at 90 °C. Samples heat-treated at 400 °C showed improved hardness thanks to crystallization and microprecipitation of Ni_3P hard phases, whereas hardness decrease was observed after treatment at 600 °C due to the combined effect of grain growth and coarsening of Ni_3P precipitates. No through-the-thickness cracks were detected by the Ferroxyl reagent after heat treatments.

Keywords: electroless Ni-P; electroless deposition; microhardness; annealing; plating parameters

Citation: Genova, V.; Paglia, L.; Pulci, G.; Pedrizzetti, G.; Pranzetti, A.; Romanelli, M.; Marra, F. Medium and High Phosphorous Ni-P Coatings Obtained via an Electroless Approach: Optimization of Solution Formulation and Characterization of Coatings. *Coatings* **2023**, *13*, 1490. https://doi.org/10.3390/coatings13091490

Academic Editor: Gergana Alexieva

Received: 17 July 2023
Revised: 17 August 2023
Accepted: 21 August 2023
Published: 24 August 2023

1. Introduction

Electroless deposition is an easy and cost-effective method for metal plating that consists of the reduction of metal ions in aqueous systems without the use of external electric current [1]. The electrons involved in the reduction reaction come from the oxidation of a reducing agent in solution. The redox process and the consequent coating growth take place on a catalytic substrate immersed in the solution. This plating technique is often referred to as autocatalytic since the growing metallic coating keeps on catalyzing the reaction and allows the continuing deposition of thick layers. Compared with the electrodeposition method, electroless plating guarantees several advantages that make it a promising technique for applications in the energy production, aerospace and automotive industries [2–4]:

- It is a non-line-of-sight technique, and the coating forms on any part of the catalytic surface placed in contact with the plating solution.
- No electric field participates in the reduction process, and a uniform, conformal and homogeneous coating can be obtained on any geometry, regardless of its shape.

- The flexibility of the solution chemistry allows its tuning to investigate innovative and cheaper formulations, to meet the specific needs of the final product [5,6].

The Ni-P system, often reported as ENP (electroless nickel plating) for simplicity, represents the most widespread technology of electroless plating: the use of hypophosphite ion as the reducing agent leads to the co-deposition of phosphorus in the nickel matrix, giving a coating with excellent properties of hardness, solderability and corrosion resistance. For this reason, Ni-P coatings are widely used to protect engineering components from surface degradation when operating environments are rich in contaminants and deteriorating agents [2,7–10]. Ni-P coatings can be classified according to the amount of phosphorous in the matrix: low phosphorous (LP, 3–5 wt% of P), medium phosphorous (MP, 6–9 wt% of P) and high phosphorous (HP, 10–14 wt% of P) [11–13]. The P quantity strongly influences coating properties by modifying the microstructure of the metallic alloy: LP coatings are crystalline, MP typically shows a mixed amorphous-crystalline microstructure, and HP is amorphous [14].

The mechanism of electroless Ni-P deposition has been extensively studied [15,16]; however, mechanisms involved in the electroless deposition process are still being discussed, and many aspects of the process remain unexplored. In any of the proposed studies, the principles governing Ni deposition are established by analyzing the main constituents of the plating solution, which are:

- Nickel source: A soluble, hydrated and stable Ni^{2+} salt. The use of sulfate and acetate gives the same results in terms of coating quality, whereas using chloride negatively influences the corrosion resistance of the produced coatings [17]. Sulphate is the most used since it is generally cheaper than acetate.
- Reducing agent: Sodium hypophosphite represents the most studied and used reducing agent [1,18].
- Complexing agent/buffer solution: The introduction of organic acids or their salts have a double function: first, they guarantee a buffer action that stabilizes the pH of the solution; second, the organic base can complex the nickel ions in solution and limit their reactivity, thus increasing bath stability. The ENP process is called autocatalytic because the substrate itself catalyzes the oxidation of the reducing agent and the reduction of the metal cations present in the solution [19]. The deposition reaction usually occurs along with the formation of byproducts that make the deposition process uncontrolled, and chemical stabilization is necessary to avoid the decomposition of the ENP plating bath. In most commercial ENP solutions, the extended service life of the bath is guaranteed by lead addition in ppm [20]. Despite this strategy being remarkably effective, the presence of Pb^{2+} in the solution causes health hazards and high costs the waste-solution disposal. The study of environmentally friendly alternatives to lead stabilizers is nowadays an important goal to improve ENP applicability and meet the strict environmental regulations towards a more sustainable development. This work presents the formulation, the study and the characterization of lead-free solutions capable of depositing medium phosphorus coatings and their optimization to increase the amount of co-deposited phosphorus to obtain HP coatings. Changes in the chemistry of the plating solution make it possible to manufacture Ni-P coatings with different P content in the alloy, according to the properties required for the specific applications. As an example, Table 1 reports different $Ni^{2+}/H_2PO_2^-$ molar ratios and the consequent wt% of P in the matrix obtained from different recipes presented in a list of works from 1994 to 2020. It can be noted that a similar molar ratio of nickel sulfate and sodium hypophosphite does not have an obvious relation to the content of P inside the matrix. This is because the P quantity does not depend only on the $Ni^{2+}/H_2PO_2^-$ ratio but on the whole system employed for the deposition, including temperature, stabilizers and complexing agents. Although all these parameters have been extensively studied, to the authors' knowledge, there are currently no works in literature that systematically relate the concentrations of the various reagents to the properties of the coatings. The tailoring of the P content by the change in concentra-

tion of reagents in the deposition solution can be a desirable aspect to increase the flexibility and applicability of specific electroless Ni-P baths. Moreover, a comprehensive understanding of the cross-relations between reagents is necessary for a better comprehension of their effect on deposition and coating properties.

Table 1. Examples of different $Ni^{2+}/H_2PO_2^-$ molar ratios used to produce a coating with different P content. --- means that the wt% *p*-values were not indicated.

	Abrantes et al. (1994) [15]	Keping and Fang (1997) [21]	Lin and Hwang (2002) [22]	Chen et al. (2002) [23]	Cheong et al. (2004) [23]	Baskaran et al. (2005) [24]	Liu et al. (2008) [25]	Rahimi et al. (2009) [26]	Wu et al. (2019) [27]	Park and Kim (2019) [28]	Lin and Chou (2020) [29]
$Ni^{2+}/H_2PO_2^-$	0.88	0.82	2.06	0.36	0.60	0.35	0.39	0.44	0.35	0.56	0.89
wt% P	12	---	8	10	10	11	10.5	9	12	3–6	5–8

For this reason, the first part of this work is focused on the investigation and correlation of different concentrations of reagents in order to methodically study their influence on P content and plating rate. Optimization of bath chemistry and deposition parameters was performed to define a recipe that can be applied for the deposition of both MP and HP coatings. Isothermal heat treatments at 400 °C and 600 °C were performed in the air in the produced samples to assess properties modification induced by the microstructure evolution. The stability and durability of the plating solution are guaranteed using an organosulfur compound as a stabilizer, and the obtained coatings were characterized in terms of composition, morphology, hardness and adhesion in the as-deposited condition and after annealing at 400 °C and 600 °C in air. The long-term goal is to provide a safe and environmentally friendly ENP solution for the deposition of MP and HP coatings that can eventually be considered a valid alternative to the more widespread lead-based plating baths.

2. Materials and Methods

2.1. Surface Preparation

All the depositions were carried out on 15 × 15 × 3 mm samples of ASTM 182 F22 steel. The procedures for surface pre-treatment and cleaning follow the B 183-79, B 322-99 and B 733-97 ASTM international standards. These procedures are necessary for removing tarnish, light rust, grease, contaminants and oxides before immersion of the specimen in the electroless plating solution. The substrates to be coated can be cleaned from grease with an alkaline solution, while the activation of the metal (i.e., removal of surface oxides that hamper deposition) is usually carried out in acidic baths. Adequate cleaning is fundamental to guarantee activation of the deposition process, and the steps of the substrate preparation procedure were defined and optimized to obtain adequate deposition conditions:

- Pre-cleaning: Performed to remove oils, lubricants or massive oxidation products that can hinder the deposition by reducing the exposed autocatalytic surface. The effect of soak cleaning and/or sandblasting was investigated. In particular, soak cleaning was carried out in a 1 M NaOH solution at 80 °C for 10 min.
- Sandblasting was used to remove the residuals from the soaked cleaning and to increase the roughness of the surface; it was performed by using corundum (mesh 80) as abrasive material.
- Acid pickling: The effect of pickling with HCl 37% at different concentrations (30, 40, 50 and 100 vol.% of HCl 37 wt%) was studied to evaluate the best activation solution. The tests were carried out on samples in the as-received condition (AR). Heat treatments in the air at 400 °C and 600 °C for 4 h were carried out to obtain a thermally grown oxide layer and simulate a surface that is not intrinsically active to electroless deposition and evaluate the effectiveness of the different activation solutions. Acid pickling was performed by immersing specimens in 100 mL of acid solution at room temperature for 1 min.

- Water rinse: After every step of the activation procedure, samples were rinsed with deionized water (DW) in an ultrasonic bath for 1 min to remove every contaminant or residue.

The success of the activation procedure was evaluated by submerging specifically pre-treated samples in the plating solution and observing the appearance of H_2 evolution (effervescence) that witnesses the proceeding of the electroless reaction. The effectiveness of activation was evaluated by assessing the deposition rate after different surface pre-treatment procedures, with higher rates determining the best results.

2.2. Solution and Coating Preparation

All chemicals were purchased from Alfa Aesar (Thermofisher Scientific, Kandel, Germany) and used without any further purification. Two recipes were optimized to obtain Ni-P coatings with different P concentrations (namely, MP and HP). All the solutions were prepared with the same procedure: the reagents were added separately and in a specific order in deionized water, and each reagent was added only after the complete dissolution of the previous one. Sodium hypophosphite was added as the first reagent; then, sodium acetate was added to stabilize the pH. Citric acid was added to set up the coordination of the Ni^{2+} that is added last as nickel sulfate. The thio-organic compound was added in the final step of the preparation procedure. Once the solution was complete, the reactor was covered and heated on a hot plate (IKA™ (Stauten, Germany) RET Control-Visc. equipped with an external temperature sensor PT1000 for continuous monitoring of the bath temperature.

The plate uses a PID (proportional–integral–derivative), controller able to maintain the target temperature with a precision of about 0.5 °C. This value is important to keep the solution stable during the deposition time. The solution is kept in constant mechanical agitation using an overhead stirrer Heidolph Hei-TORQUE Core (Heidolph Instruments GmbH & Co. KG, Schwabach, Germany) equipped with PTFE coated impeller. pH was constantly monitored using a METTLER TOLEDOTM Seven Excellence pH-meter model S400, equipped with pH sensor InLab® Viscous Pro-ISM (Mettler Toledo, Columbus, OH, USA). Temperature, agitation and pH value were continuously controlled during the deposition. A recipe for the production of MP coating was first optimized (with formulation reported in Table 2); after that, the quantities of the nickel source ($NiSO_4$) and the complexing agent (citric acid) were modified in order to obtain a formulation for HP coatings. The properties of the coatings obtained with the modified formulation were analyzed considering both the variation in the concentration of a single reagent and the combined effect of the concentration change of both the nickel source and the complexing agent. The buffer quantity (sodium acetate) has not been changed in order to maintain the pH constant at a value of 4.2. It is well known that pH strongly influences the P content and the plating rate since it induces a shift in the equilibrium of reactions, and its effect was extensively studied in the literature [1]; variation of the pH values implies the use of a different buffer agent that would invariably influence the solution formulation. For this reason, the buffer reagent and its concentration were excluded from the studied parameters.

Table 2. Medium phosphorous (MP) bath formulation.

Function	Name	Chemical Formula	MP (g/L)
Reducing agent	Sodium hypophosphite	NaH_2PO_2	70
Buffer	Sodium acetate	$C_2H_3NaO_2$	15
Chelating Agent	Citric acid	$C_6H_8O_7$	7
Source of Nickel	Nickel sulfate	$NiSO_4$	12
Stabilizer *	Thio-organic compound (TOC)	R-CS	5 (ppm)

* The stabilizer was added by liquid solution (1 mol/kg) and respecting the quantity in ppm.

The results regarding the quality of the coating are presented in terms of plating rate (thickness measured by cross-section SEM analysis) and composition. The P content in the coating was measured by EDS on cross-section micrographs. Analysis was performed on at least five areas comprising 70% of the coating thickness, starting from the top of the coating. Despite the narrow sensitivity of the instrument, EDS was selected as a technique to determine the P quantity as reported in several works, because it is widely recognized as a technique for determining the amount of phosphorus in an electroless Ni-P system [15,22,24,27,28,30,31]. All the coated samples, unless otherwise stated, were thermally treated (right after the deposition) at 180 °C for 2 h in air to avoid hydrogen embrittlement phenomena [17,32,33]: such coatings will be referred to as as-coated hereafter. Isothermal heat treatments at 400 °C and 600 °C for 1 h were carried out on all the produced samples in order to study the evolution of properties upon crystallization. Variation of crystallite size with thermal treatments was calculated using Scherrer's equation [34]:

$$D = \frac{0.94\lambda}{\beta \cos(\theta)}, \quad (1)$$

where λ is the wavelength of the radiation used, β is the half-maximum width, and θ is the position of the main peak. No correction for instrumental broadening was made.

All treatments were performed in the air in a Lenton tube furnace (now Carbolite Gero Ltd., Sheffield, UK) with a heating rate of 10° C/min.

2.3. Coating Characterization

Surface morphology, thickness and composition of the coatings were determined with a FEG-SEM Tescan Mira3 (Tescan, Brno, Czech Republic) equipped with Edax Octane Elect EDS system detector (Edax/Ametek Inc. Pleasanton, CA, USA) for energy-dispersive X-Ray spectroscopy (EDS); Edax Team v.4.5 software was employed for the elementary analysis. The top view observation is important to evaluate changes in the surface morphology and microstructure according to the different deposition parameters, while the observation of cross-sections allows measurement of the coating thickness, evaluation of the deposition rate and assessment of the P content (expressed in wt%) through coating thickness. Samples were prepared for cross-sectional analysis by mounting in epoxy resin, with subsequent cut with a slow-speed precision saw and polished with SiC papers (grit P400 to P1200) and diamond suspensions (up to 1 μm). X-ray diffraction analysis (XRD) was performed with a Philips X'Pert device (PANalytical B.V., Almelo, The Netherlands) to identify the crystalline phases in the coatings. The XRD device operated at 40 kV and 40 mA with CuKα1 radiation, a scan range of 30–80° (2θ), a step size of 0.02° and a counting time of 2 s.

Mechanical characterization tests were performed on the optimized MP and HP coatings in the as-coated condition and after heat treatments at 400 °C and 600 °C. Vickers microhardness was measured according to ASTM E384-11 using a LEICA VMHT (Leica GmbH, Wetzlar, Germany) equipped with a Vickers diamond indenter (load 50 gf, time 15 s). The distance between two indentations was $\geq$50 μm, and results report the average and standard deviation of at least fifteen measurements for each coating. Adhesion strength of coatings was measured with an Instron 5584 (Illinois Tool Works Inc., Norwood, MA, USA) mechanical testing machine equipped with a load cell of 150 kN at a crosshead rate of 2 mm/min according to the ASTM C633. Each specimen was obtained by assembling one cylindrical coated substrate to the sandblasted faces of the loading fixtures by an adhesive bonding agent (Polyamide-epoxy FM 1000 Adhesive Film, CyTech, Treviso, Italy); a self-aligning device was adopted to guarantee a perfect co-axiality of the two parts during the assembly and the heat treatment necessary for the cure of the glue (180 °C for 1 h).

The presence of through-the-thickness porosities or cracks was detected by the ferroxyl reagent test according to the procedure of ASTM B689-87: a filter paper was placed on the sample, and some drops of the ferroxyl solution were poured on it with a Pasteur pipette. The ferroxyl solution consists of 10 g/L potassium hexacyanoferrate (III), 60 g/L sodium chloride, 20 mL of phenolphthalein for 1 L of deionized water. The solution turns blue when Fe^{2+} ions are detected: if blue spots appear on the filter paper, it means that the solution reached the surface of the steel substrate, and the test can be considered positive, revealing the presence of through-the-thickness porosities or cracks.

3. Results

3.1. Surface Preparation

The formation of a thin oxide layer on the surface of the specimens can hind the activation of the deposition reaction since the substrate is required to be conductive to trigger the oxidation reaction. For this reason, specific and effective surface pre-treatments are important steps to avoid inhomogeneities in the first layer of the coating. The effect of surface pre-deposition treatments (referred to as pre-treatments) was studied by varying the conditions of acid pickling (concentration of HCl 37% was varied from 30 vol.% to 100 vol.%), and results are reported in Table 3. Pre-treatments were performed on received (AR) substrates and on heat-treated substrates (at 400 and 600 °C for 1 h in air). Samples are reported as "Activated" whether the deposition was successful, and, in such cases, the deposition rates are reported in the last column of Table 3. When tests are performed on AR substrates, two cases do not guarantee activation: when 10 min lab air exposure follows the pre-cleaning process (soak cleaning in NaOH and sandblasting) and when acid pickling is performed with 100 vol.% of HCl 37 wt%. In the first case, the pre-cleaning effectively degreases the substrate, and sandblasting removes initial oxides, but ferrous substrates exhibit low corrosion resistance, and 10 min of lab air exposure is sufficient to cause further oxidation and hinder deposition. In the case of HCl 100 vol.%, the acid treatment for pickling can be considered too aggressive and leads to the formation of corrosion products that inhibit the deposition process [35]. This last phenomenon, known as over-corrosion, can sometimes be observed when substrates are cleaned by pickling with aggressive acids [36]. The reactions involved in the acid pickling process can be summarized in the following reactions:

$$Fe_3O_4(s) + 8H^+(l) \rightarrow 2Fe^{3+}(l) + Fe^{2+} + 4H_2O(l) \quad (2)$$

$$Fe_2O_3(s) + 6H^+(l) \rightarrow 2Fe^{3+}(l) + 3H_2O(l) \quad (3)$$

$$FeO(s) + 2H^+(l) \rightarrow Fe^{2+}(l) + H_2O(l) \quad (4)$$

$$Fe(s) + 2H^+(l) \rightarrow Fe^{2+}(l) + H_2(g) \quad (5)$$

Equations (2)–(4) refer to the removal of oxides from the metal surface, while Equation (5) is related to the phenomenon of over-corrosion, which causes the dissolution of Fe and the release of Fe ions in solution. If the pickling process is performed using an excessively strong acid, like HCl 100 vol.%, the removal of the oxides that naturally grow on the surface is so fast that some of the pickling solution reacts with the base metal, making it more prone to oxidation when exposed to air. The XRD analysis before and after the activation treatments was performed to better investigate the formation of products that prevent effective activation of the substrate. Figure 1 shows a comparison of XRD spectra of the as-received substrate (i) before the removal of oil and lubricants that prevent oxidation during the storage, (ii) after the pre-cleaning process (soaking in NaOH 1 M at 80 °C for 1 h and sandblasting) and exposure to 10 min lab air, (iii) after treatment with HCl 100 vol.% and exposure to 10 min lab air and (iv) after treatment with HCl 40 vol.% and exposure to 10 min lab air. All samples were protected with Cortec® corrosion inhibitor oil before XRD analysis to prevent further air exposure. Peaks at 2θ position 45° and 65° are referred to as Fe (JCPDS no. 87-0722). As expected, only Fe peaks are present in the AR samples. Considering the pre-treated substrates, both the AR samples after pre-cleaning and after acid pickling with HCl 100% show Fe_3O_4 peak (magnetite, JCPDS no. 76-0955), which confirms the presence of an inert oxide layer that prevents activation of the deposition. Conversely, no magnetite peaks are detected on the XRD spectrum of the sample treated with HCl 40 vol.%. This suggests that both sandblasting and over-corrosion make the substrate more vulnerable to oxidation, probably due to the larger exposed surface, whereas samples that underwent pickling with HCl 40 vol.% require a longer amount of time to oxidate and show the best results in terms of activation when immersed in the plating solution.

Table 3. Activation results after acid pickling. AR refers to the samples in the as-received condition; HT400C refers to the samples oxidized at 400 °C for 4 h; HT600C refers to the samples oxidized at 600 °C for 4 h.

Sample	HCl Concentration (%vol)	Result	Plating Rate (mg/cm^2/h)
AR	(Pre-cleaning and 1 h air exposure, no activation procedure)	Not Activated	-
	30	Activated	8.60
	40	Activated	15.7
	50	Activated	14.0
	100	Not activated	-
HT 400 °C	(Pre-cleaning and 1 h air exposure, no activation procedure)	Not Activated	-
	30	Activated	13.8
	40	Activated	12.9
	50	Activated	15.1
	100	Activated	11.0
HT 600 °C	(Pre-cleaning and 1 h air exposure, no activation procedure)	Not Activated	-
	30	Not activated	-
	40	Not activated	-
	50	Not Activated	-
	100	Activated	9.80

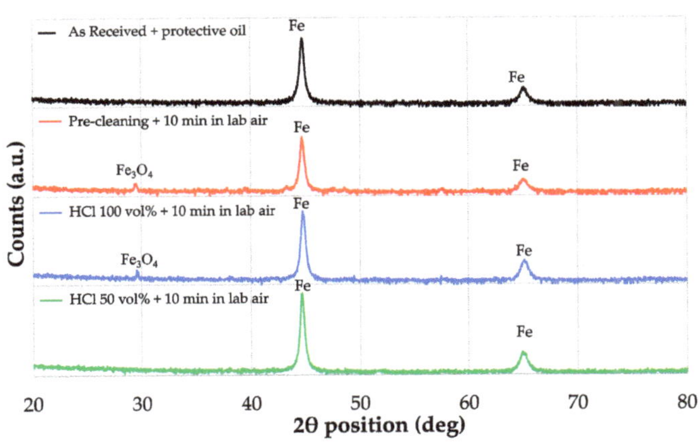

Figure 1. Comparison of XRD spectra of as received sample (black), after pre-cleaning and 10 min lab air exposure (red), after acid picking with HCl 100 vol.% and 10 min lab air exposure (blue) and after acid picking with HCl 50 vol.% and 10 min lab air exposure (green).

Therefore, the success of the activation procedure strongly depends on the balance between the strength of the acid for pickling and the amount of oxide on the surface of the substrate (the degradation of the substrate and the amount of the oxide on the surface strongly depends on the preservation and storage conditions of the components). To better understand this phenomenon, the substrates were oxidized by exposure at 400 °C and 600 °C for 4 h in the air using a Lenton tube furnace (now Carbolite Gero Ltd., Sheffield, UK) to simulate different conditions of degradation. The specimens treated at 400 °C are expected to exhibit a more pronounced surface oxidation than the AR, and results in Table 3 show that successful activation can be obtained for all concentrations of HCl. Conversely, the substrates treated at 600 °C are oxidized to the extent that only pickling with undiluted HCl can provide activation. The strength of the acidic solutions used for the activation treatment also influences the plating rate: the slower deposition associated with lower concentrations of HCl can be addressed to incomplete pickling, which is only capable of removing some part of the oxide scale. This leads to smaller exposure of the catalytic surface and a lower number of nucleation sites catalytically active for deposition. In such cases, the Ni-P coating mainly forms by lateral growth of the present nucleation germs and the plating rate is globally slower. The best results are therefore obtained when achieving a good balance between the HCl concentration of the acidic solution and the amount of oxide to be removed.

The XRD analysis of the activation tests on the substrates heat-treated at 400 °C and 600 °C are reported in Figure 2a and 2b, respectively. The spectra of the samples just after heat treatment reveal that the high temperate exposure leads to the formation of Fe_2O_3 (hematite, JCPDS no. 73-2234). Either the pre-cleaning process (here given by sandblasting only) and the acid pickling with HCl 100% are sufficient to remove the oxide grown upon thermal treatments, so hematite is no more detectable from XRD; nevertheless, just like in the case of AR samples, sandblasting and acid pickling with HCl 100 vol.% expose the base metal, causing the formation of a layer of magnetite layer when exposed to air. However, the activation tests demonstrate that the amount of over-corrosion is not sufficient to totally prevent activation and deposition proceeds at a slower rate.

Figure 3 shows the schematic representation of the acidic pickling process with the two extremes where activation is not obtained: the "soft acid" refers to the case in which the pickling does not remove the oxide scale from the substrate so that the catalytic surface is not exposed and the deposition is hampered; the "hard acid" case occurs when the base metal is over-corroded, and the surface is more vulnerable to oxidation, causing formation of products that can hinder (or slow down) deposition.

According to the experimental activity, the optimal combination of soak cleaning, sandblasting and acid pickling was selected as the standard activation procedure and reported in Table 4.

Table 4. Standardized procedure of surface pre-deposition activation treatment.

Treatment	Time
NaOH-1 M-80 °C	10 min
Water rinsing	-
Sandblasting	-
Sonication in H_2O	5 min
HCl 50 %vol	1 min
Water rinsing	-
Plating	-

The sonication step in deionized water (DW) was added since it is necessary to remove the abrasive material that remains stuck in the substrate after sandblasting.

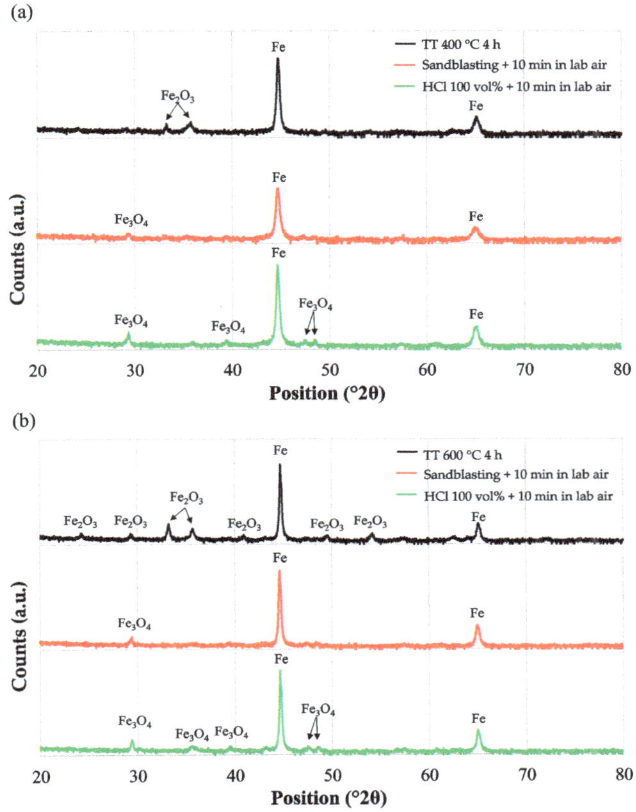

Figure 2. (**a**) XRD spectra of samples oxidized at 400 °C for 4 h: without any treatment (top), after sandblasting + air lab exposure of 1 h (center) and after acid pickling with HCl 100 vol.% (bottom) (**b**) XRD spectra of samples oxidized at 600 °C for 4 h: without any treatment (top), after sandblasting + air lab exposure of 1 h (center) and after acid pickling with HCl 100 vol.% (bottom).

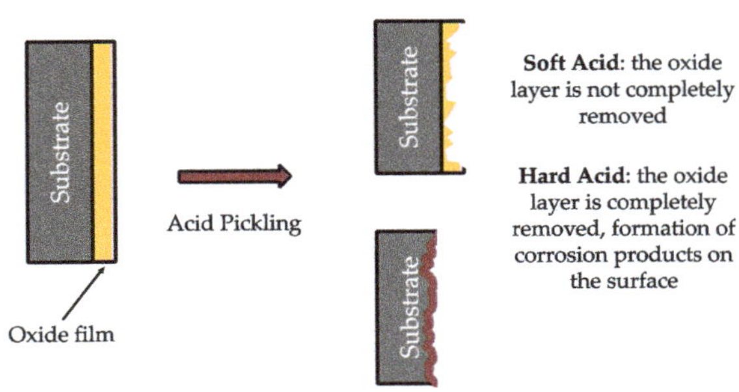

Figure 3. Acid pickling effects the oxide layer in the case where activation is not achieved.

3.2. Study of the Solution

The influence of temperature on the deposition rate was investigated for the MP plating solution (Figure 4). Only temperatures that guarantee a deposition rate higher than 5 μm/h are reported in the graph, but this limit does not represent the lower temperate of reaction activation. Acidic baths with comparable formulations are reported in the literature to be active at temperatures as low as 50 °C [37], but too-slow deposition rates may limit applicability and increase costs when thick functional coatings are required (e.g., anti-wear applications). Temperatures higher than 95 °C have also been excluded to avoid significant evaporation of the plating solution and the consequent concentration of the reagents. EDS analysis confirmed that the plating temperature does not influence the %P for MP deposition, as similarly reported in other studies [35]. Considering these results, the selected operating temperature was 90 °C.

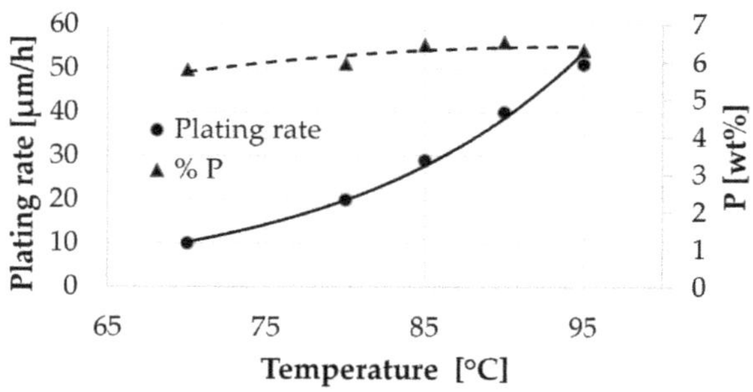

Figure 4. Plating rate and %P as a temperature function for MP plating solution. Dashed and solid lines are inserted for the readability of the graph and do not represent any mathematical relation.

Given the stability and high plating rate of MP formulation, the concentration of reagents was optimized in order to increase the amount of co-deposited phosphorus, with the final aim of producing HP coatings.

The first studied reagent was $NiSO_4$ since it represents the source of nickel-free ions in solution. Reducing its quantity, without any variation on the concentration of the complexing agent and the reducing agent, theoretically entails the decrease of deposition rate due to the lower Ni^{2+} availability [38], according to the reaction reported in Equation (6) [31]:

$$Ni^{2+} + 2H_2PO_2^- + 2H_2O \rightarrow Ni^0 + 2H_2PO_3^- + 2H^+ + H_2 \quad (6)$$

This trend is confirmed by the results in Figure 5, where the plating rate was evaluated after reducing $NiSO_4$ quantity by 15, 30, 45 and 60 wt%. from the standard MP formulation. Furthermore, EDS analysis on these coatings confirms that the decrease of $NiSO_4$ in solution is also associated with an increase of co-deposited P [16]. This result suggests that the availability of free nickel ions is a key factor in increasing the P content in the Ni-P alloy; however, it is also associated with a dramatic reduction in the plating rate.

Another strategy for obtaining a Ni-P coating with a higher P quantity consists in increasing the concentration of the complexing agent. Complexation is known to reduce the number of free nickel ions in the plating solution and change the mixed current potential, leading to both a reduction in the plating rate and changes in coating composition [39]. The effect of citric acid amount, with a fixed concentration of nickel sulfate and hypophosphite, was studied by increasing MP standard formulation by 10 wt%, 20 wt% and 30 wt%. The results in Figure 6 show that higher amounts of citric acid associate with a slower plating rate, only up to 20 wt% increase, where equi-molarity with Ni^{2+} ions is achieved (i.e., all the citrate ions are bound to all Ni^{2+} ions) [40]. Beyond this value, an increase in the plating rate is registered, and it can mainly be ascribed to the ΔpH induced by the presence of citric acid that is not bound to nickel ions [41]. A different trend is observed for P content, which is lower than the standard MP formulation for a citric acid increase of 10 wt% and 20 wt%. This is due to the increased chelation of Ni^{2+} ions, which stabilizes the oxidized form of the metal and hinders its reduction, limiting the whole redox process [42]. When +30 wt% of citric acid is added, the lower pH promotes the deposition process and P deposition [1]. The addition of +40% of citric acid was tested, but the excessive increase in the complexing agent concentration results in a large amount of adsorption onto the catalytic surface, with a poisoning effect that totally inhibits the deposition process [41,43].

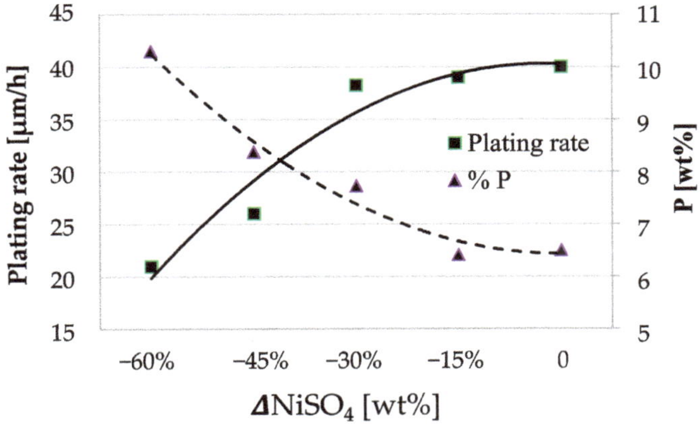

Figure 5. Variation of plating rate and P wt% as a function of $\Delta NiSO_4$ from MP formulation. Operating temperature: 90 °C. Dashed and solid lines are inserted for the readability of the graph and do not represent any mathematical relation.

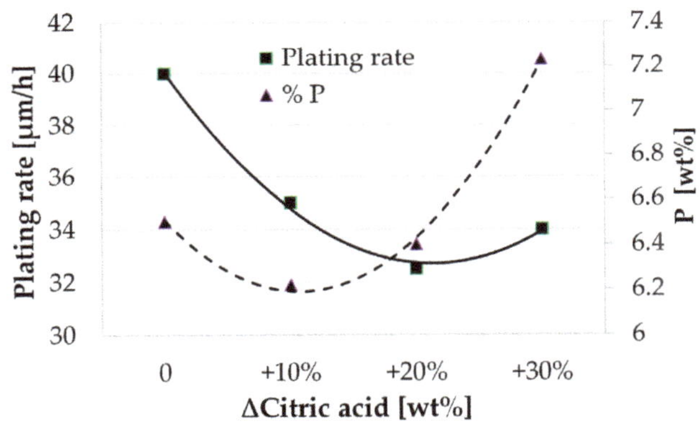

Figure 6. Plating rate and P wt% as a function of ΔCitric acid from standard MP plating solution. Operating temperature = 90° C. Dashed, and solid lines are inserted for the readability of the graph and do not represent any mathematical relation.

Despite the promising result on P content increasing upon +30 wt% of citric acid, ~7 wt% of P is not sufficient to classify the coating as HP. For this reason, simultaneous variations of $NiSO_4$ and citric acid were studied to consider the cross-correlation between the two parameters. Variations from the MP standard formulation will be addressed with the nomenclature $MP_{x|y}$, where ±x represents the variation of $NiSO_4$ and ±y is the variation of citric acid. Figure 7 shows the results in a 3D histogram chart (mean results of 5 depositions of each experimental point are reported, the calculated standard deviation is always lower than 3% and is not reported in the figure for readability). The empty spots represent a combination for which the deposition cannot be activated; this can be traced back to the combined effect of decreasing availability of Ni^{2+} ions when $NiSO_4$ concentration is decreased and the increased complexation upon the increasing amount of citric acid. The formulation with +10 wt% of citric acid and −45 wt% of $NiSO_4$ provides the best results in terms of plating rate (~30 μm/h) and %P (10.5 wt%). This confirms that the influence of reagents cannot be treated separately and that properties of the system strongly depend on the mutual influence of every component: the amount of $NiSO_4$ has a strong influence on the amount of co-deposited P, but a good compromise with plating rate and bath stability can only be achieved by the concurrent optimization of the quantity of the complexing agent.

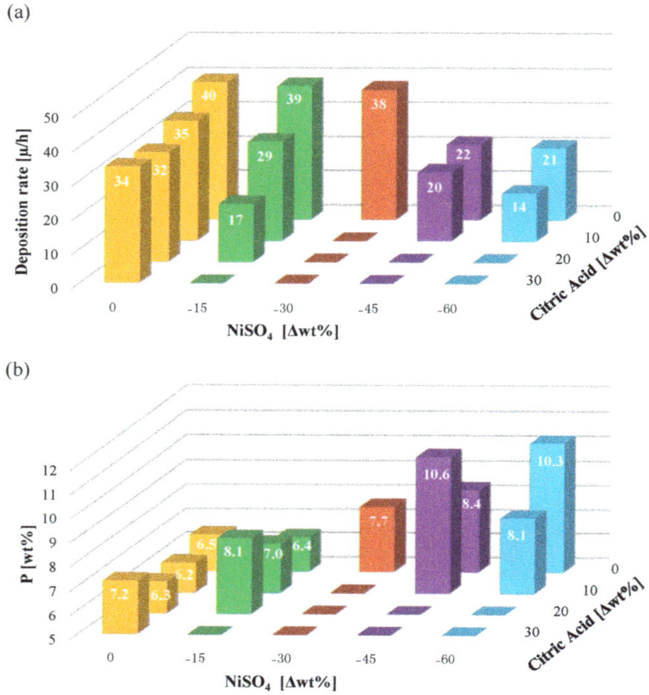

Figure 7. Plating rate and %P as a function of $\Delta NiSO_4$ and a Δ Citric acid with respect to MP solution formulation. Operating temperature = 90 °C.

It should be remarked that the presented step of the experimental activity was conducted by keeping the concentration of the reducing agent constant, giving a variation in the $[Ni^{2+}]/[H_2PO_2^-]$ addressed to the change in $NiSO_4$ concentration. To uncover the correlation between P wt% and the concentration of hypophosphite ions, the molar ratio $[Ni^{2+}]/[H_2PO_2^-]$ was varied by changing the amount of the reducing agent, while all other reagents were kept constant. Concentrations of the hypophosphite ion were modified in order to vary the molar ratio $[Ni^{2+}]/[H_2PO_2^-]$ in the range 1/25–1/3, and experimental activity was carried out on three formulations of the plating solution: (i) the standard MP formulation ($MP_{0|0}$), (ii) $MP_{-30|0}$, which shows intermediate properties of %P and plating rate, and (iii) $MP_{-45|+10}$, which guarantees the higher amount of co-deposited P.

Curves of %P variation are reported in Figure 8, and it can be noted that changing hypophosphite concentration leads to a non-linear variation of the amount of co-deposited P, with a similar trend for all the investigated formulations. This phenomenon can be traced by considering the reactions involved in the Ni-P deposition system [16]:

$$H_2PO_2^- + H_2O \rightarrow H_2PO_3^- + 2H^+ + 2e^- \quad (7)$$

$$Ni^{2+} + 2e^- \rightarrow Ni \quad (8)$$

$$H_2PO_2^- + 2H^+ + e^- \rightarrow P + 2H_2O \quad (9)$$

$$2H^{+} + 2e^{-} \rightarrow H_2 \tag{10}$$

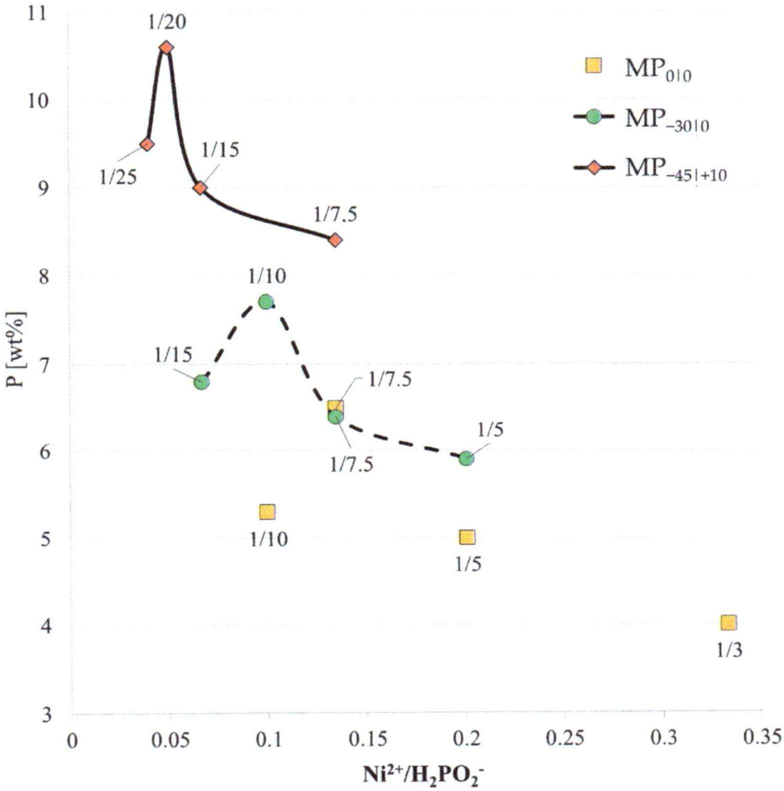

Figure 8. Phosphorous percentage (%P) as a function of $[Ni^{2+}]/[H_2PO_2^{-}]$. Operating temperature = 90 °C.

Reactions in Equations (7) and (8) refer, respectively, to the oxidation of hypophosphite and to the reduction of nickel; the reaction in Equation (9) describes the mechanism of phosphorus formation and its co-deposition into the alloy, whereas Equation (10) is a side reaction that leads to hydrogen evolution and makes no contribution to the deposition. The formation of the Ni-P matrix, given by Ni deposition and P co-deposition, can be considered a multi-step mechanism of parallel reactions that are interdependent on one another since reactions (7) and (8) are combined and (7) and (9) are combined. Even though a full understanding of the kinetics of the system goes beyond the scope of this work, Figure 8 depicts that P wt% in the coatings strongly depends on $Ni^{2+}/H_2PO_2^{-}$ ratio. Each curve was obtained with fixed Ni^{2+} concentration and represented %P co-deposition according to $H_2PO_2^{-}$ variation. The amount of phosphorus in the coating with sodium hypophosphite could be illustrated according to the empirical kinetics equation [16,44]:

$$\frac{d[P]}{dt} = k\left[H_2PO_2^{-}\right]^{1.91}\left[H^{+}\right]^{0.25} \tag{11}$$

The amount of P in the coating increases with increasing concentration of hypophosphite ions, as could be expected and consistently with Equation (11), but only up to a threshold where the maximum of the curves is reached. A further increase in the quantity of $H_2PO_2^-$ in the plating bath leads to lower amounts of co-deposited P. A similar behavior was observed by Sun et al. [44], who attributed this effect to the intensification of hypophosphite diffusion and adsorption, together with the increase in the proportion of hydrogen evolution. Each curve in Figure 8 follows a similar trend, with a maximum that is reached for progressively higher amounts of hypophosphite ions for formulation with a lower quantity of Ni^{2+}. Moreover, the maximum of %P is lower for higher concentrations of nickel in the plating solution, in accordance with the findings about the [$NiSO_4$] effect presented above. This last result confirms that the influence of each reagent cannot be evaluated separately from the rest of the system. The presented curves can represent a useful tool for the final optimization of the P wt% in the coating, according to the required properties given by the application.

It is worth mentioning that changes in the reducing agent concentration do not dramatically influence the deposition rate, as shown in Figure 9, and the main contribution is given by Ni^{2+} availability.

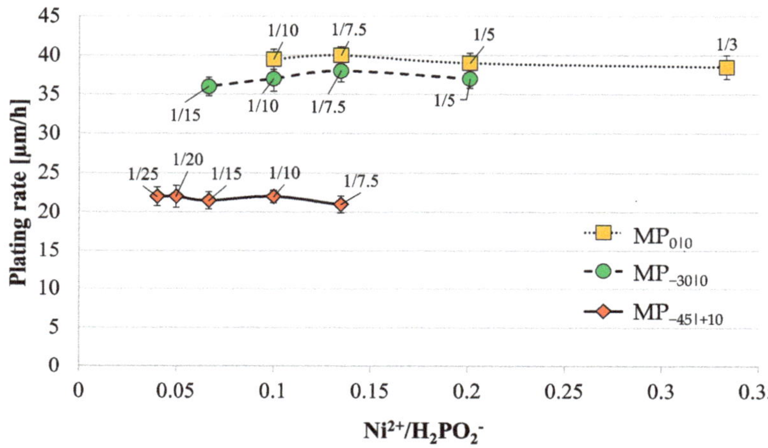

Figure 9. Deposition rate as a function of $[Ni^{2+}]/[H_2PO_2^-]$. Operating temperature = 90 °C.

The formulation $MP_{-45|+10|}$ with a molar ratio $[Ni^{2+}]/[H_2PO_2^-]$ equals 1/20 was selected as the best candidate for the HP formulation. However, it is poorly stable with respect to the standard MP solution, and adequate service life of the plating bath cannot be guaranteed. Therefore, the influence of increasing the concentration of the stabilizer (i.e., the thio-organic compound, or TOC) was studied to optimize its concentration in the HP solution. Results showing the plating rate and the wt% of P as a function of TOC concentration are reported in Figure 10. As similarly reported in the literature [16], the addition of the stabilizer does not dramatically affect the deposition rate, although a beneficial effect can be obtained when the concentration rises up to 8.5 ppm. Bath stability is guaranteed for concentrations in the range of 5–9 ppm: below 5 ppm, the solution is characterized by low stability and degrades at a temperature close to 90 °C, foreclosing the attainment of the best temperature for deposition; conversely concentrations higher than 9 ppm completely hamper the deposition to start. The last phenomenon is well-studied in the literature [30,45] and known as the "poisoning mechanism": sulfur from the thio-group of the TOC adsorbs on catalytically active sites to control deposition reactions, but when in excessive quantity, it can completely inhibit the process. Below the threshold of 5 ppm, the concentration of the stabilizer is too low to efficiently control the deposition; as the concentration increases, progressively fewer catalytic sites will be active for deposition, inhibiting the reduction of metal ions on undesired surfaces with high surface energy, like particles in solution or catalytic sites in the reactor/container, thus preventing random bath decomposition. A more efficient deposition on the substrate is therefore achieved and associated with the increase in plating rate [24]. Nevertheless, when the concentration limit is exceeded, the poisoning mechanism takes place on the substrate itself so that it is no longer available as a catalytic surface, and the reaction does not start. The concentration of the stabilizer is also known to influence the P content in the deposit, leading to its increase when the cathodic reduction of nickel is inhibited [20,24,31]. It is important to point out that deposition with a TOC concentration of 5 ppm is characterized by a high plating rate and high P wt% but low stability that might lead to random bath decomposition. Therefore, 8.5 ppm was selected as the best quantity for the formulation of HP solution in order to ensure great stability of the plating bath, high plating rate and high wt% of P.

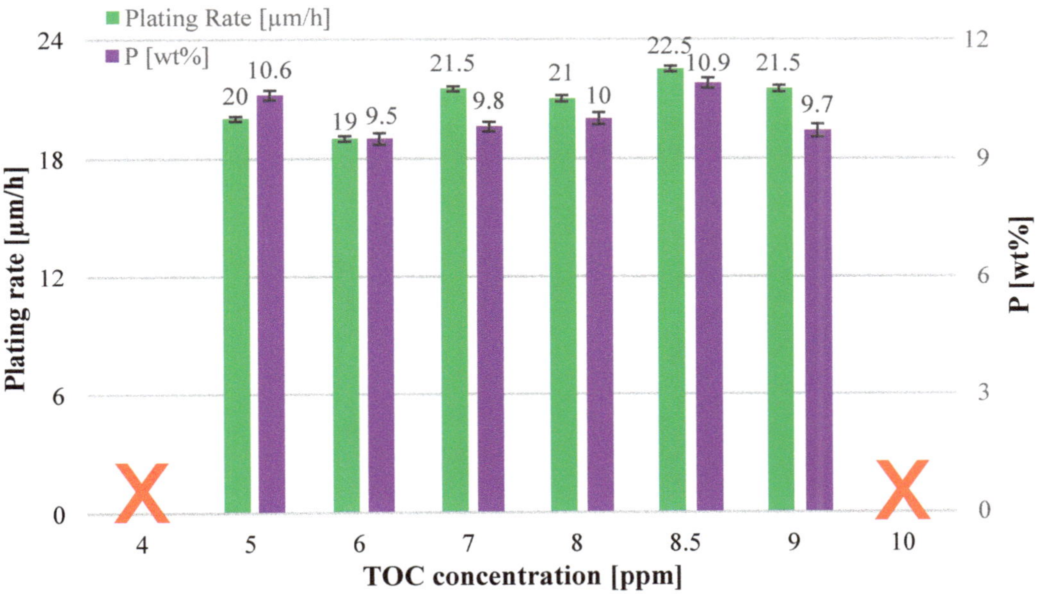

Figure 10. Plating rate and %P as a function of ppm of thio-organic compound. Plating temperature = 90 °C. $MP_{45.10\text{–}1/20}$ formulation.

To conclude, the study of all deposition parameters eventually led to the formulation of a stable solution for the deposition of HP coatings (10.9 wt% of P) and the composition is reported in Table 5.

Table 5. High phosphorous (HP) bath formulation obtained after optimization of the process parameters.

Function	Name	Chemical Formula	HP (g/L)
Reducing agent	Sodium hypophosphite	NaH_2PO_2	51.5
Buffer	Sodium acetate	$C_2H_3NaO_2$	15.0
Chelating Agent	Citric acid	$C_6H_8O_7$	7.70
Source of Nickel	Nickel sulfate	$NiSO_4$	6.60
Stabilizer *	Thio-organic compound (TOC)	R-CS	8.50 (ppm)

* The stabilizer was added by liquid solution (1 mol/kg) in accordance with the quantity in ppm.

3.3. Coating Microstructure and Composition

The SEM surface micrographs of the MP and HP coatings are reported in Figure 11a,b, corresponding to the XRD microstructural analysis and EDS elemental analysis presented in Figure 11c and 11d, respectively. Both samples exhibit a nodular morphology with a cauliflower-like structure, typical of coatings obtained by the electroless plating technique [46,47]; however, HP coatings show a denser network with a smaller size of individual nodules than the MP. This effect can be addressed by the higher quantity of stabilizer in the HP solution, which limits lateral growth [24,31,48]. The comparison of the two XRD patterns shows that MP coatings are characterized by a higher degree of crystallinity compared to HP ones. These characteristics are expected because the higher P content in the Ni matrix distorts the lattice to an extent where amorphous nickel is obtained [49], whereas P content in MP coatings is only sufficient to refine grain size and create a nanocrystalline structure [14]. Yet, the amorphous nature of HP coatings is responsible for their high corrosion resistance [50,51]. All the peaks in the XRD spectra (45°, 52° and 75°) refer to cubic Ni (JCPDS 65-2865) (Figure 11). The use of thio-stabilizers is sometimes associated with S contamination in the coatings, with negative effects on corrosion resistance [52]. Nevertheless, the EDS spectrum reveals the presence of Ni and P only, confirming the purity of the obtained coatings. The quantitative EDS analysis performed on the top view of the two coatings shows content of P equal to 10.9 ± 0.3 wt% for the HP and 6.5 ± 0.2 wt% for the MP.

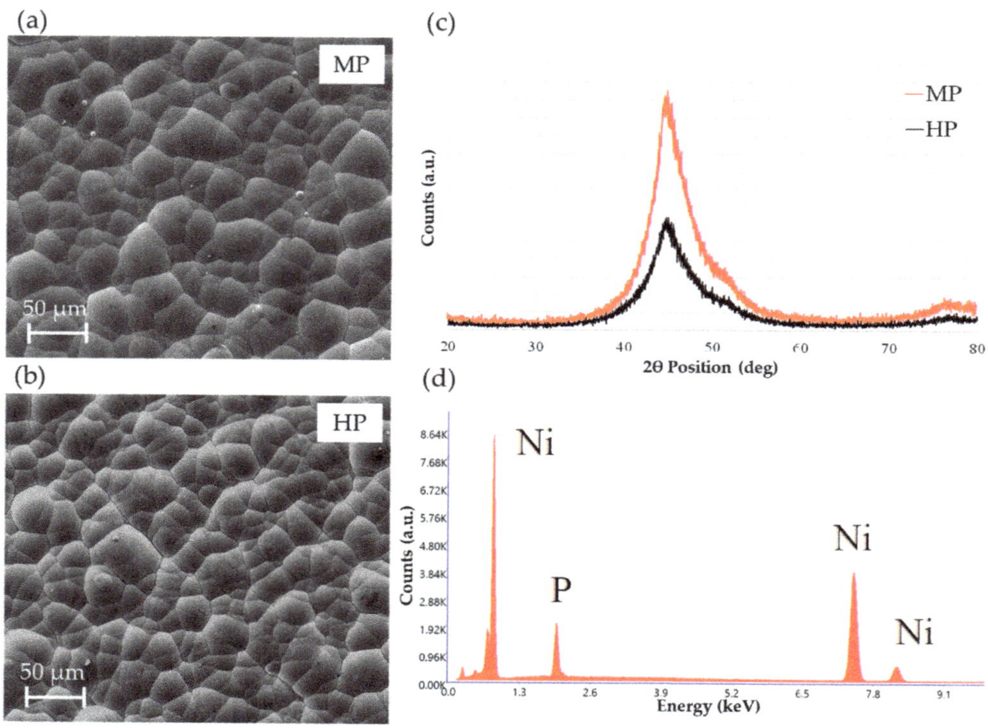

Figure 11. Morphological and elementary analysis for MP (6.5 wt%) and HP (10.9 wt%) coatings obtained by top-view SEM images (**a**,**b** respectively), X-ray diffraction (**c**) and EDS (**d**).

3.4. *Pull Off-Test*

Pull-off tests were performed to compare the adhesion strength of MP and HP coatings obtained after simple pre-cleaning (soaking in NaOH and sandblasting) and after the standard pre-treatment process (pre-cleaning and activation by pickling with HCl 50%, as defined in Section 3.1). During an adhesion test, detachment between the coated rod and the uncoated one (counterpart) can happen by different mechanisms: (i) detachment between the glue and the counterpart; (ii) detachment between the glue and the coated road; (iii) detachment between the coating and the substrate; (iv) cohesive failure of the coating. These different kinds of rupture are not necessarily independent and might also occur simultaneously [53]. The tested samples coated after the standard pre-treatment always experienced a cohesive/adhesive separation between the glue and the counterpart, as shown in Figure 12a, whereas samples coated just after pre-cleaning typically showed rupture by a partial detachment of the coating (Figure 12b).

(a) (b)

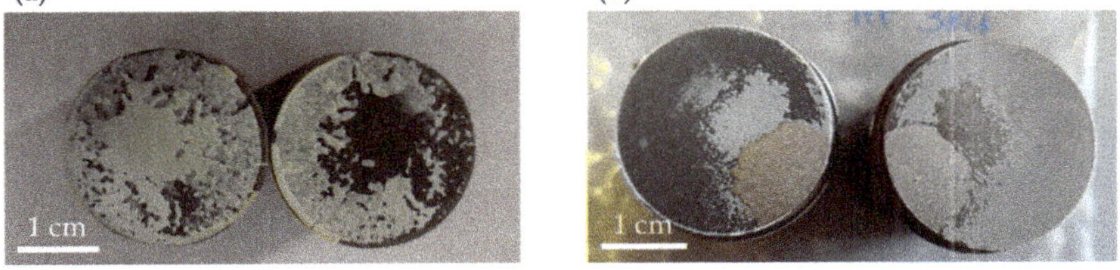

Figure 12. Example of adhesive/cohesive glue detachment in pre-treated samples (**a**) and partial detachment of the coating when no pre-treatment was performed (**b**).

Accounting that the reference value for the bonding strength of the adhesive film was 80 ± 1 MPa, the values of the maximum load experienced by the samples before rupture are reassumed in Figure 13. Both MP and HP coatings obtained after standard activation pre-treatment are characterized by better adhesion compared to those that underwent pre-cleaning only. Moreover, the samples coated without the activation procedure are characterized by lower adhesion and exhibit failure by a partial detachment of the coating from the substrate, probably due to the incomplete removal of oxides or products from long-term storage of substrates. Surface activation is known to have an important role in adhesion strength between the coating and the substrate [54,55] since they affect the mechanism of nucleation and growth of the Ni-P deposit. The presence of impurities, such as greases, sandblasting residues and /or oxide layers, may favor the formation of pores and cracks at the coating-substrate interface, thus decreasing overall adhesion. Therefore, both results about the maximum load before rupture and analysis of the failure mechanism enlighten the better adhesion of the coating when a standard activation pre-treatment is carried out.

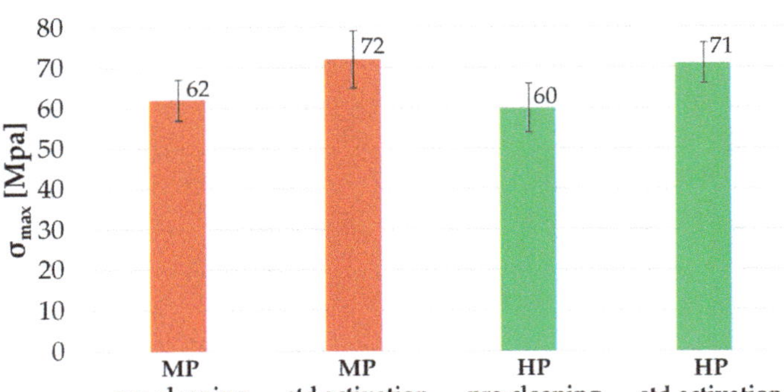

Figure 13. Pull-off test for the MP and HP samples obtained with standard activation treatments compared with those obtained with pre-cleaning only. The reference value of the adhesive film is 80 MPa.

3.5. Heat Treatments

Heat treatments above the crystallization temperature of the Ni-P alloy (reported to be around 340 °C [34,56,57]) are often employed to modify coating properties; therefore, the evolution of coating characteristics upon high-temperature exposure at 400 °C and 600 °C for 1 h in the air was investigated. All the samples involved in the characterization process are summarized in Table 6.

Table 6. Samples (MP and HP) for morphological, microstructural and mechanical characterization.

Sample	wt% P	Heat Treatment	Activation
MP_{AC}	6.5	-	HCl 50 vol.%, 1 min
MP_{400}	6.5	400 °C, 1 h	HCl 50 vol.%, 1 min
MP_{600}	6.5	600 °C, 1 h	HCl 50 vol.%, 1 min
HP_{AC}	10.9	-	HCl 50 vol.%, 1 min
HP_{400}	10.9	400 °C, 1 h	HCl 50 vol.%, 1 min
HP_{600}	10.9	600 °C, 1 h	HCl 50 vol.%, 1 min

SEM micrographs in Figure 14 show the morphological evolution of samples according to heat treatment temperature. Surface morphology of MP coatings in the as-deposited condition, after annealing for 1 h at 400 °C and 600 °C are reported in Figure 14a, 14b and 14c, respectively; the same surface condition for HP coatings are reported in Figure 14d–f, respectively. It can be observed that the thermal treatment at 400 °C does not induce particular morphological modifications; conversely, exposure at 600 °C leads to changes in morphology and an oxide scale visibly begins to grow on the surface.

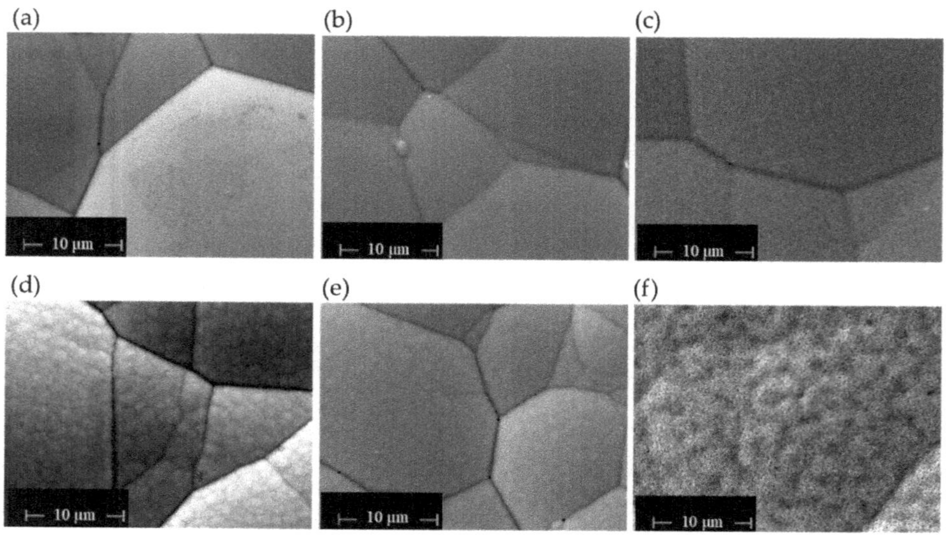

Figure 14. Morphological analysis (SEM top-view) for (**a**) MP as deposited; (**b**) MP after annealing at 400 °C for 1 h in air; (**c**) MP after annealing at 600 °C for 1 h in air; (**d**) HP as deposited; (**e**) HP after annealing at 400 °C for 1 h in air; (**f**) HP after annealing at 600 °C for 1 h in air.

XRD spectra in Figure 15 show the microstructural evolution of MP (a) and HP (b) coatings as a function of heat treatment temperature. Substantial changes in microstructure are observed in both MP and HP coatings after treatment at 400 °C: recrystallization of the f.c.c. Ni matrix is observed together with precipitation of b.c.t. Ni_3P [49,58–62]. A segregation process of phosphorus at the grain boundaries and triple junctions of the Ni-

P grains occurs with increasing temperature, leading to the formation of P-rich regions [46] and precipitation of Ni_3P when P concentration exceeds a certain threshold (reported to be around 15 wt% of P) [63]. There is evidence that the Ni_3P precipitation preferentially occurs at grain boundaries and triple junctions [64,65]; therefore, MP coatings (nanocrystalline in the as-deposited condition) experience a continuous phosphorus segregation process when temperature increases. However, Farber et al. [46] proposed that structural changes upon heat treatments depend on the formation of a metastable grain boundary phase that already develops in the as-deposited state of HP coatings. For this reason, precipitation of the Ni_3P compound is thought to have taken place earlier in the HP deposits [34,66], and a massive presence of precipitates can be observed on the HP spectrum, even though they are characterized by lower crystallinity than MP coatings. In fact, sharper Ni peaks can be observed for MP, which associates with a greater grain size of the matrix. After treatment at 600 °C, both spectra become sharper, indicating grain growth of both Ni and Ni_3P phases. NiO peaks also appear, confirming the presence of a thin scale of oxide, as observed by SEM images. The finer morphology of HP coatings and their lower surface roughness are expected to influence the growth of the thin oxide film and can also be responsible for the different morphology of the scale. The evolution of average grain size, calculated with the Scherrer equation (Equation (1)), and coarsening of Ni_3P precipitates with increasing temperature is reported in Figure 16.

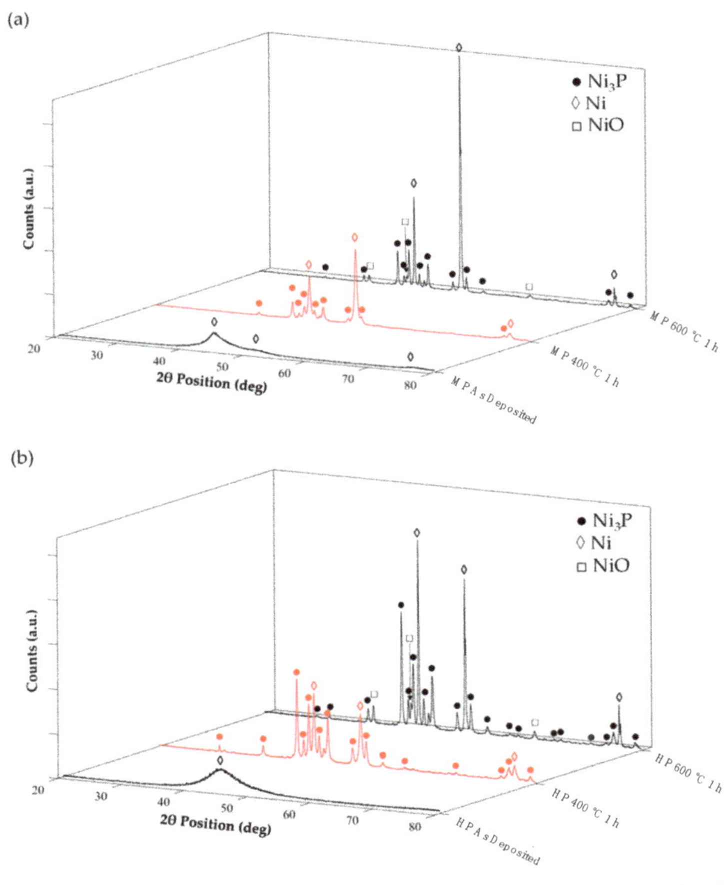

Figure 15. XRD spectra for MP (**a**) and HP (**b**) coatings showing microstructural evolution upon heat treatments for 1 h in air.

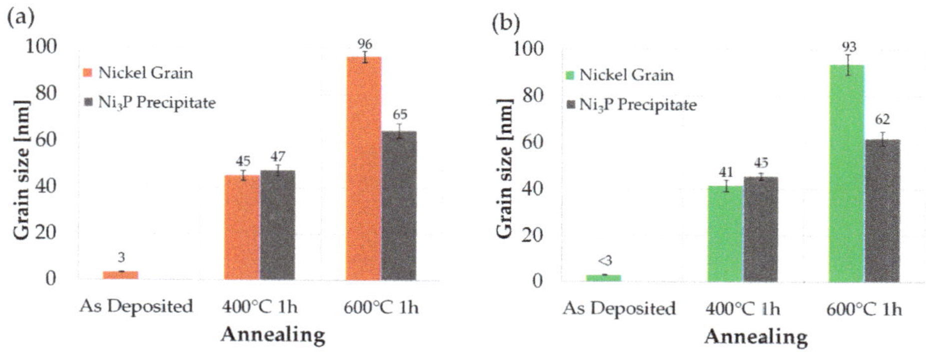

Figure 16. Increase of average nickel crystallite size and coarsening of Ni_3P precipitates with increasing annealing temperature for MP (**a**) and HP (**b**) coatings.

No trace of superficial or internal cracks induced by the heat treatment was observed by SEM analysis, and samples were tested negative for the ferroxyl reagent test.

3.6. Vickers Hardness

The microhardness of MP and HP coatings in the as-coated condition and after heat treatments are shown in Figure 17. The nanocrystalline structure of deposited MP coatings guarantees higher hardness than amorphous HP. As shown in Figure 16, heat treatment at 400 °C is associated with increased crystallite size of nickel and precipitation of hard Ni_3P phases, which lead to hardness increase of the deposits. In contrast, a dramatic hardness decrease is registered after treatment at 600 °C due to excessive grain growth and the formation of coarse precipitates. Deformation processes of deposited Ni-P coatings, which have very fine grain size or can even be amorphous, show a reverse Hall-Patch effect [67]; therefore, grain growth induced by heat treatment at 400 °C associates with lower grain boundary sliding and rotation [68,69] and results in increased microhardness. The presented results are in accordance with literature data [49,60,61,70,71] that report maximum hardness after heat treatments at 400 °C where Ni_3P precipitates guarantee a precipitation hardening effect and grain growth is limited to an extent that causes strengthening. Conversely, the samples treated at 600 °C exhibit lower microhardness values because of the excessive grain growth and the formation of coarse precipitates with a less effective hardening effect in accordance with the Orowan strengthening mechanism [72].

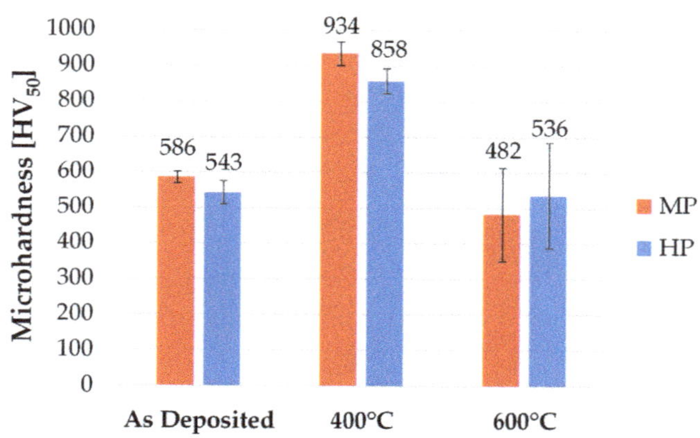

Figure 17. Microhardness (HV) for the MP and HP coatings in the as-deposited condition and after heat treatment at 400 °C and 600 °C for 1 h in air.

4. Conclusions

A new lead-free electroless nickel plating process was developed for the plating of MP coatings and optimized to obtain a formulation for HP deposition. Pre-deposition treatments are found to be crucial for the activation of deposition and in determining the adhesion strength of the coatings, with the best results of plating rate obtained when performing a pre-treatment of pickling with HCl 50%vol for 1 min. A study on the chemical composition of the solution demonstrated that the ratio between the reducing agent and metal ion is a key parameter for the tailoring of the P co-deposition. In particular, decreasing the $Ni^{2+}/H_2PO_2^-$ ratio by a reduction of 40 wt% of $NiSO_4$ and increasing of 10 wt% of complexing agent leads P co-deposition up to 10.6 wt%. Further optimization of the TOC allows the P content to rise up to 10.8 wt%, while guaranteeing the best results on bath stability. Best deposition parameters guaranteed a plating rate of 40 μm/h for the MP coatings and 25 μm/h for HP coatings, with deposition at 90 °C. Characterization of coatings confirmed the nanocrystalline and amorphous nature of MP and HP coatings, respectively, with hardness up to 930 HV and 840 HV when heat-treated at 400 °C for 1 h. No through-thickness cracks were detected by the ferroxyl reagent tests performed on the samples treated at 400 °C.

Author Contributions: Conceptualization, A.P., M.R. and F.M.; validation, G.P. (Giovanni Pulci); formal analysis, V.G. and L.P.; investigation, V.G., L.P. and G.P. (Giulia Pedrizzetti); data curation, G.P. (Giulia Pedrizzetti); writing—original draft, V.G.; writing—review and editing, G.P. (Giovanni Pulci) and G.P. (Giulia Pedrizzetti); supervision, G.P. (Giovanni Pulci) and F.M.; project administration, F.M.; funding acquisition, A.P. and M.R. All authors have read and agreed to the published version of the manuscript.

Funding: This research received no external funding.

Institutional Review Board Statement: Not applicable.

Informed Consent Statement: Not applicable.

Data Availability Statement: Not applicable.

Conflicts of Interest: The authors declare no conflict of interest.

References

1. Loto, C.A. Electroless Nickel Plating—A Review. *Silicon* **2016**, *8*, 177–186. [CrossRef]
2. Sahoo, P.; Das, S.K. Tribology of Electroless Nickel Coatings—A Review. *Mater. Des.* **2011**, *32*, 1760–1775. [CrossRef]
3. Pedrizzetti, G.; Paglia, L.; Genova, V.; Conti, M.; Baiamonte, L.; Marra, F. The Effect of Composition and Heat Treatment on Microhardness of Ni-P and Ni-P-NanoZrO_2 Coatings. *Chem. Eng. Trans.* **2023**, *100*, 433–438. [CrossRef]
4. Genova, V.; Pedrizzetti, G.; Paglia, L.; Marra, F.; Bartuli, C.; Pulci, G. Diffusion Aluminide Coating Modified via Electroless Nickel Plating for Ni-Based Superalloy Protection. *Surf. Coat. Technol.* **2022**, *439*, 128452. [CrossRef]
5. Hari Krishnan, K.; John, S.; Srinivasan, K.N.; Praveen, J.; Ganesan, M.; Kavimani, P.M. An Overall Aspect of Electroless Ni-P Depositions-A Review Article. *Metall. Mater. Trans. A* **2006**, *37A*, 1917–1926. [CrossRef]
6. Genova, V.; Paglia, L.; Marra, F.; Bartuli, C.; Pulci, G. Pure Thick Nickel Coating Obtained by Electroless Plating: Surface Characterization and Wetting Properties. *Surf. Coat. Technol.* **2019**, *357*, 595–603. [CrossRef]
7. Singh, R. The Application of an Electrochemical Technique to Determine the Porosity of Electroless Nickel Coatings Produced in Hypophosphite Baths. *Anti-Corros. Methods Mater.* **2003**, *50*, 121–128. [CrossRef]
8. Lelevic, A.; Walsh, F.C. Electrodeposition of Ni-P Alloy Coatings: A Review. *Surf. Coat. Technol.* **2019**, *369*, 198–220. [CrossRef]
9. Sankara Narayanan, T.S.N.; Baskaran, I.; Krishnaveni, K.; Parthiban, S. Deposition of Electroless Ni-P Graded Coatings and Evaluation of Their Corrosion Resistance. *Surf. Coat. Technol.* **2006**, *200*, 3438–3445. [CrossRef]
10. Dhakal, D.R.; Kshetri, Y.K.; Chaudhary, B.; Kim, T.H.; Lee, S.W.; Kim, B.S.; Song, Y.; Kim, H.S.; Kim, H.H. Particle-Size-Dependent Anticorrosion Performance of the Si3 N4-Nanoparticle-Incorporated Electroless Ni-P Coating. *Coatings* **2022**, *12*, 9. [CrossRef]
11. Gould, A.J.; Boden, P.J.; Harris, S.J. Phosphorus Distribution in Electroless Nickel Deposits. *Surf. Technol.* **1981**, *12*, 93–102. [CrossRef]
12. Keong, K.G.; Sha, W.; Malinov, S. Crystallisation Kinetics and Phase Transformation Behaviour of Electroless Nickel-Phosphorus Deposits with High Phosphorus Content. *J. Alloys Compd.* **2002**, *334*, 192–199. [CrossRef]
13. Buchtík, M.; Kosár, P.; Wasserbauer, J.; Tkacz, J.; Doležal, P. Characterization of Electroless Ni-P Coating Prepared on a Wrought ZE10 Magnesium Alloy. *Coatings* **2018**, *8*, 96. [CrossRef]

14. Agarwala, R.C.; Agarwala, V. Electroless Alloy/Composite Coatings: A Review. *Sadhana* **2003**, *28*, 475–493. [CrossRef]
15. Abrantes, L.M.; Correia, J.P. On the Mechanism of Electroless Ni-P Plating. *J. Electrochem. Soc.* **1994**, *141*, 2356. [CrossRef]
16. Mallory, G.O.; Hajdu, J.B. *Electroless Plating: Fundamentals and Applications*; American Electroplaters and Surface Finishers Society: Orlando, FL, USA, 1990.
17. Inam-Ul-Haque; Ahmad, S.; Khan, A. Electroless Nickel Plating on ABS Plastics from Nickel and Nickel Sulfate Baths. *J. Chem. Soc. Pak.* **2005**, *27*, 246–249.
18. Stankiewicz, A.; Szczygieł, I.; Szczygieł, B. Summary of Existing Models of the Ni-P Coating Electroless Deposition Process. *Int. J. Chem. Kinet.* **2013**, *45*, 755–762. [CrossRef]
19. Genova, V.; Marini, D.; Valente, M.; Marra, F.; Pulci, G. Nanostructured Nickel Film Deposition on Carbon Fibers for Improving Reinforcement-Matrix Interface in Metal Matrix Composites. *Chem. Eng. Trans.* **2017**, *60*, 73–78. [CrossRef]
20. Das, L.; Chin, D.-T. Effect of Bath Stabilizers on Electroless Nickel Deposition on Ferrous Substrates. *Plat. Surf. Finish.* **1996**, *83*, 55–61.
21. Keping, H.; Fang, J.L. Stabilization Effect of Electroless Nickel Plating by Thiourea. *Met. Finish.* **1997**, *95*, 73–75. [CrossRef]
22. Lin, K.-L.; Hwang, J.-W. Effect of Thiourea and Lead Acetate on the Deposition of Electroless Nickel. *Mater. Chem. Phys.* **2002**, *76*, 204–211. [CrossRef]
23. Chen, B.H.; Hong, L.; Ma, Y.; Ko, T.M. Effects of Surfactants in an Electroless Nickel-Plating Bath on the Properties of Ni-P Alloy Deposits. *Ind. Eng. Chem. Res.* **2002**, *41*, 2668–2678. [CrossRef]
24. Baskaran, I.; Sankara Narayanan, T.S.N.; Stephen, A. Effect of Accelerators and Stabilizers on the Formation and Characteristics of Electroless Ni-P Deposits. *Mater. Chem. Phys.* **2006**, *99*, 117–126. [CrossRef]
25. Liu, H.P.; Li, N.; Bi, S.F.; Li, D.Y.; Zou, Z.L. Effect of Organic Additives on the Corrosion Resistance Properties of Electroless Nickel Deposits. *Thin Solid Films* **2008**, *516*, 1883–1889. [CrossRef]
26. Rahimi, A.R.; Modarres, H.; Abdouss, M. Study on Morphology and Corrosion Resistance of Electroless Ni-P Coatings. *Surf. Eng.* **2009**, *25*, 367–371. [CrossRef]
27. Wu, W.; Liu, J.; Miao, N.; Jiang, J.; Zhang, Y.; Zhang, L.; Yuan, N.; Wang, Q.; Tang, L. Influence of Thiourea on Electroless Ni–P Films Deposited on Silicon Substrates. *J. Mater. Sci. Mater. Electron.* **2019**, *30*, 7717–7724. [CrossRef]
28. Park, I.C.; Kim, S.J. Effect of Lead Nitrate Concentration on Electroless Nickel Plating Characteristics of Gray Cast Iron. *Surf. Coat. Technol.* **2019**, *376*, 2–7. [CrossRef]
29. Lin, J.D.; Chou, C. Te the Influence of Phosphorus Content on the Microstructure and Specific Capacitance of Etched Electroless Ni-P Coatings. *Surf. Coat. Technol.* **2019**, *368*, 126–137. [CrossRef]
30. Cheong, W.J.; Luan, B.L.; Shoesmith, D.W. Protective Coating on Mg AZ91D Alloy—The Effect of Electroless Nickel (EN) Bath Stabilizers on Corrosion Behaviour of Ni-P Deposit. *Corros. Sci.* **2007**, *49*, 1777–1798. [CrossRef]
31. Cheong, W.J.; Luan, B.L.; Shoesmith, D.W. The Effects of Stabilizers on the Bath Stability of Electroless Ni Deposition and the Deposit. *Appl. Surf. Sci.* **2004**, *229*, 282–300. [CrossRef]
32. Tulsi, S.S. Properties of Electroless Nickel. *Trans. Inst. Met. Finish.* **1986**, *64*, 73–76. [CrossRef]
33. Staia, M.H.; Castillo, E.J.; Puchi, E.S.; Lewis, B.; Hintermann, H.E. Wear Performance and Mechanism of Electroless Ni-P Coating. *Surf. Coat. Technol.* **1996**, *86*, 598–602. [CrossRef]
34. Apachitei, I.; Tichelaar, F.D.; Duszczyk, J.; Katgerman, L. The Effect of Heat Treatment on the Structure and Abrasive Wear Resistance of Autocatalytic NiP and NiP-SiC Coatings. *Surf. Coat. Technol.* **2002**, *149*, 263–278. [CrossRef]
35. Sudagar, J.; Lian, J.; Sha, W. Electroless Nickel, Alloy, Composite and Nano Coatings—A Critical Review. *J. Alloys Compd.* **2013**, *571*, 183–204. [CrossRef]
36. Tang, B.; Li, D.; Fu, F.; Xu, Y.; Yu, G.; Zhang, J. A Strategy for Cleaner Pickling: Effect, Mechanism, and Evaluation Method of a Complex-Inhibitor in Hydrochloric Acid Medium. *Ind. Eng. Chem. Res.* **2012**, *51*, 2615–2621. [CrossRef]
37. Huang, Z.; Nguyen, T.T.; Zhou, Y.; Qi, G. A Low Temperature Electroless Nickel Plating Chemistry. *Surf. Coat. Technol.* **2019**, *372*, 160–165. [CrossRef]
38. Salicio-Paz, A.; Ugarte, I.; Sort, J.; Pellicer, E.; García-Lecina, E. Full Optimization of an Electroless Nickel Solution: Boosting the Performance of Low-Phosphorous Coatings. *Materials* **2021**, *14*, 1501. [CrossRef]
39. Huang, Y.S.; Cui, F.Z. Effect of Complexing Agent on the Morphology and Microstructure of Electroless Deposited Ni–P Alloy. *Surf. Coat. Technol.* **2007**, *201*, 5416–5418. [CrossRef]
40. Baker, E.N.; Baker, H.M.; Anderson, B.F.; Reeves, R.D. Chelation of Nickel (II) by Citrate. The Crystal Structure of a Nickel-Citrate. *Inorganica Chim. Acta* **1983**, *78*, 281–285. [CrossRef]
41. Jin, Y.; Yu, H.; Yu, D.; Sun, D. Effects of Complexing Agents on Acidic Electroless Nickel Deposition. *Rare Met.* **2010**, *29*, 401–406. [CrossRef]
42. Bard, A.J.; Faulkner, L.R.; White, H.S. *Electrochemical Methods: Fundamentals and Applications*; Wiley: New York, NY, USA, 1980; ISBN 0-471-05542-5.
43. Lin, Y.-M.; Yen, S.-C. Effects of Additives and Chelating Agents on Electroless Copper Plating. *Appl. Surf. Sci.* **2001**, *178*, 116–126. [CrossRef]
44. Sun, R.; Yu, G.; Xie, Z.; Hu, B.; Zhang, J.; He, X.; Zhang, X. Influence of Hypophosphite on Efficiency and Coating Qualities of Electroless Ni-P Deposits on Magnesium Alloy AZ91D. *Int. J. Electrochem. Sci.* **2015**, *10*, 7893–7904. [CrossRef]
45. Feldstein, N.; Amodio, P.R. Anionic Inhibition in Electroless Plating. *J. Electrochem. Soc.* **1970**, *117*, 1110. [CrossRef]

46. Farber, B.; Cadel, E.; Menand, A.; Schmitz, G.; Kirchheim, R. Phosphorus Segregation in Nanocrystalline Ni-3.6 At% Alloy Investigated with the Tomographic Atom Probe (TAP). *Acta Mater.* **2000**, *48*, 789–796. [CrossRef]
47. Elsener, B.; Crobu, M.; Scorciapino, M.A.; Rossi, A. Electroless Deposited Ni-P Alloys: Corrosion Resistance Mechanism. *J. Appl. Electrochem.* **2008**, *38*, 1053–1060. [CrossRef]
48. Azumi, K.; Yugiri, T.; Kurihara, T.; Seo, M.; Habazaki, H.; Fujimoto, S. Direct Plating of Electroless Ni-P Layers on Sputter-Deposited Al-Ni Alloy Films. *J. Electrochem. Soc.* **2003**, *150*, C461. [CrossRef]
49. Keong, K.G.; Sha, W.; Malinov, S. Hardness Evolution of Electroless Nickel-Phosphorus Deposits with Thermal Processing. *Surf. Coat. Technol.* **2003**, *168*, 263–274. [CrossRef]
50. Fayyad, E.M.; Abdullah, A.M.; Hassan, M.K.; Mohamed, A.M.; Jarjoura, G.; Farhat, Z. Recent Advances in Electroless-Plated Ni-P and Its Composites for Erosion and Corrosion Applications: A Review. *Emergent Mater.* **2018**, *1*, 3–24. [CrossRef]
51. Crobu, M.; Scorciapino, A.; Elsener, B.; Rossi, A. The Corrosion Resistance of Electroless Deposited Nano-Crystalline Ni-P Alloys. *Electrochim. Acta* **2008**, *53*, 3364–3370. [CrossRef]
52. Wu, C.Y.; Chen, Y.H.; Tang, Y.K.; Lin, E.J.; Lin, Y.X.; Wang, J.Y.; Zhuang, W.X.; Lee, C.H.; Chiu, C.Y.; Yeh, C.Y.; et al. Effect of Chemical Additives in the Plating Bath on Surface Corrosion Resistance of Ni(P). *J. Electron. Mater.* **2020**, *49*, 26–33. [CrossRef]
53. Valli, J. A Review of Adhesion Test Methods for Thin Hard Coatings. *J. Vac. Sci. Technol. A Vac. Surf. Film.* **1986**, *4*, 3007–3014. [CrossRef]
54. Correa, E.; Zuleta, A.A.; Sepúlveda, M.; Guerra, L.; Castaño, J.G.; Echeverría, F.; Liu, H.; Skeldon, P.; Thompson, G.E. Nickel-Boron Plating on Magnesium and AZ91D Alloy by a Chromium-Free Electroless Process. *Surf. Coat. Technol.* **2012**, *206*, 3088–3093. [CrossRef]
55. Guo, C.Y.; Matinlinna, J.P.; Tin, A.; Tang, H. A Novel Effect of Sandblasting on Titanium Surface: Static Charge Generation. *J. Adhes. Sci. Technol.* **2012**, *26*, 2603–2613. [CrossRef]
56. Huang, Y.S.; Zeng, X.T.; Annergren, I.; Liu, F.M. Development of Electroless NiP-PTFE-SiC Composite Coating. *Surf. Coat. Technol.* **2003**, *167*, 207–211. [CrossRef]
57. Jiaqiang, G.; Lei, L.; Yating, W.; Bin, S.; Wenbin, H. Electroless Ni-P-SiC Composite Coatings with Superfine Particles. *Surf. Coat. Technol.* **2006**, *200*, 5836–5842. [CrossRef]
58. Franco, M.; Sha, W.; Malinov, S.; Rajendran, R. Phase Composition, Microstructure and Microhardness of Electroless Nickel Composite Coating Co-Deposited with SiC on Cast Aluminium LM24 Alloy Substrate. *Surf. Coat. Technol.* **2013**, *235*, 755–763. [CrossRef]
59. Jiang, J.; Chen, H.; Zhu, L.; Qian, W.; Han, S.; Lin, H.; Wu, H. Effect of Heat Treatment on Structures and Mechanical Properties of Electroless Ni-P-GO Composite Coatings. *RSC Adv.* **2016**, *6*, 109001–109008. [CrossRef]
60. Luo, H.; Leitch, M.; Behnamian, Y.; Ma, Y.; Zeng, H.; Luo, J.L. Development of Electroless Ni-P/Nano-WC Composite Coatings and Investigation on Its Properties. *Surf. Coat. Technol.* **2015**, *277*, 99–106. [CrossRef]
61. Kundu, S.; Das, S.K.; Sahoo, P. Properties of Electroless Nickel at Elevated Temperature—A Review. In Proceedings of the Procedia Engineering; Elsevier Ltd.: Amsterdam, The Netherlands, 2014; Volume 97, pp. 1698–1706.
62. Buchtík, M.; Krystýnová, M.; Másilko, J.; Wasserbauer, J. The Effect of Heat Treatment on Properties of Ni-P Coatings Deposited on a AZ91 Magnesium Alloy. *Coatings* **2019**, *9*, 461. [CrossRef]
63. Hentschel, T.H.; Isheim, D.; Kirchheim, R.; Mu, F.; Ller, È.; Kreye, H. Nanocrystalline Ni-3.6 at.% P and Its Transformation Sequence Studied by Atom-Probe Field-Ion Microscopy. *Acta Mater.* **2000**, *48*, 933–941. [CrossRef]
64. Boylan, K.I.; Ostrander, D.I.; Erb, U.I.; Palumbo, G.; Aust, K. An In-Situ TEM Study of the Thermal Stability of Nanocrystalline Ni-P. *Scr. Metall. Mater.* **1991**, *25*, 2711–2716. [CrossRef]
65. Kalidindi, A.R.; Schuh, C.A. Phase Transitions in Stable Nanocrystalline Alloys. *J. Mater. Res.* **2017**, *32*, 1993–2002. [CrossRef]
66. Buchtík, M.; Doskočil, L.; Brescher, R.; Doležal, P.; Másilko, J.; Wasserbauer, J. The Effect of Crystallization and Phase Transformation on the Mechanical and Electrochemical Corrosion Properties of Ni-p Coatings. *Coatings* **2021**, *11*, 447. [CrossRef]
67. Apachitei, I.; Duszczyk, J. Autocatalytic Nickel Coatings on Aluminium with Improved Abrasive Wear Resistance. *Surf. Coat. Technol.* **2000**, *132*, 8998. [CrossRef]
68. van Swygenhoven, H.; Caro, A. Plastic Behavior of Nanophase Ni: A Molecular Dynamics Computer Simulation. *Appl. Phys. Lett.* **1997**, *71*, 1652–1654. [CrossRef]
69. Johnson, C.E.; Ogburn, F. Hardness of Heat Treated Electroless Nickel. *Surf. Technol.* **1976**, *4*, 161. [CrossRef]
70. Franco, M.; Sha, W.; Aldic, G.; Malinov, S.; Çimenoğlu, H. Effect of Reinforcement and Heat Treatment on Elevated Temperature Sliding of Electroless Ni-P/SiC Composite Coatings. *Tribol. Int.* **2016**, *97*, 265–271. [CrossRef]
71. Zhang, X.; Wang, H.; Liu, G. Friction and Wear of Electroless Ni-P-CS Composite Coating. *Metals* **2023**, *13*, 315. [CrossRef]
72. Martin, J.W. *Precipitation Hardening*; Butterworth-Heinemann: Oxford, UK, 1998; ISBN 0750638850.

coatings

MDPI

Article

Rapid Thermal Processing of Kesterite Thin Films

Maxim Ganchev [1], Stanka Spasova [1,*], Taavi Raadik [2], Arvo Mere [2], Mare Altosaar [2] and Enn Mellikov [2]

1 Central Laboratory of Solar Energy and New Energy Sources, Bulgarian Academy of Sciences, 72 Tzarigradsko Shaussee Blvd., 1784 Sofia, Bulgaria; mganchev@abv.bg

2 Department of Materials Science, Tallinn University of Technology, Ehitajate tee 5, 19086 Tallinn, Estonia; taavi.raadik@taltech.ee (T.R.); arvo.mere@taltech.ee (A.M.); mare.altosaar@taltech.ee (M.A.); ennm@staff.ttu.ee (E.M.)

* Correspondence: perovskite.psc@gmail.com

Abstract: Multinary chalcogenides with Kesterite structure $Cu_2ZnSn(S,Se)_4$ (CZTSSe) are a prospective material base for the enhancement of the photovoltaics industry with abundant and environmentally friendly constituents and appropriate electro-physical properties for building highly efficient devices at a low cost with a short energy pay-back time. The actual record efficiency of 13.6%, which was reached recently, is far below the current isostructural chalcopyrite's solar cells efficiency of near 24%. The main problems for future improvements are the defects in and stability of the Kesterite absorber itself and recombination losses at interfaces at the buffer and back contacts. Here, we present an investigation into the rapid thermal annealing (RTA) of as-electrodeposited thin films of Cu_2ZnSnS_4 (CZTS). The treatment was carried out in a cold wall tubular reactor in dynamic conditions with variations in the temperature, speed and time of the specific elements of the process. The effect of annealing was investigated by X-ray diffractometry, Raman scattering and Scanning Electron Microscopy (SEM). The phase composition of the films depending on treatment conditions was analyzed, showing that, in a slow, prolonged, high-temperature process, the low-temperature binaries react completely and only Kesterite and ZnS are left. In addition, structural investigations by XRD have shown a gradual decrease in crystallite sizes when the temperature level and duration of the high-temperature segment increases, and respectively increase in the strain due to the formation of the phases in non-equilibrium conditions. However, when the speed of dynamic segments in the process decreases, both the crystallite size and strain of the Kesterite non-monotonically decrease. The grain sizes of Kesterite, presented by SEM investigations, have been shown to increase when the temperature and the duration increase, while the speed decreases, except at higher temperatures of near 750 °C. The set of experiments, following a scrupulous analysis of Raman data, were shown to have the potential to elucidate a way to ensure the fine manipulation of the substitutional Cu/Zn defects in the structure of CZTS thin films, considering the dependences of the ratios of $Q = I_{287}/I_{303}$ and $Q' = I_{338}/(I_{366} + I_{374})$ on the process variables. Qualitatively, it can be concluded that increases in the speed, duration and temperature of RTA lead to increases in the order of the structure, whereas, at higher temperatures of near 750 °C, these factors decrease.

Keywords: photovoltaics; multinary chalcogenides; thermal annealing; Kesterite; XRD

Citation: Ganchev, M.; Spasova, S.; Raadik, T.; Mere, A.; Altosaar, M.; Mellikov, E. Rapid Thermal Processing of Kesterite Thin Films. *Coatings* **2023**, *13*, 1449. https://doi.org/10.3390/coatings13081449

Academic Editor: Aomar Hadjadj

Received: 27 June 2023
Revised: 8 August 2023
Accepted: 10 August 2023
Published: 17 August 2023

1. Introduction

The increased capacity of photovoltaics worldwide is could lead to the achievement of an ecological footprint of near-zero by the year 2030. Actually, silicon-based solar cell technology dominates the photovoltaics market, comprising near 95% of all installed modules [1]. Multinary chalcogenides are large class of compounds with special applications in photovoltaics. Beginning with chalcopyrite $CuInSe_2$, $Cu(In,Ga)Se_2$, and $CuInS_2$ (CIGS), a sustainable generation of compounds was established, successfully competing with silicon-based photovoltaic technologies [2,3] with record efficiencies of 23.4% for the Cd-free CIGS device [4] and 26.2% for CIGS-based solar cells with an electron back-reflector [5]. They have several advantages, such as a sub-micrometer thickness, very high absorption coefficient near 10^5 cm^{-1} and tunable bandgap from 1.0 to 1.5 eV depending on the kind and ratio of chalcogens.

An interesting approach has been developed involving the formation of bifacial solar cells growing the structure with a $Cu(In,Ga)Se_2$ absorber on glass or a flexible substrate [6]. Recently, a record bifacial power conversion efficiency was demonstrated of near 20% and near 11% under frontal and rear illumination. The power generation density of the device was predicted to be comparable to the record obtained for the mono-facial option.

These factors encourage mass production but the scarcity of rare-earth elements in Ga limits the their potential application. In an attempt to overcome this obstacle, a new class of compounds of Kesterite was developed where the couple In and Ga is replaced with abundant elements Zn and Sn. Similar to Chalcopyrite (CIGSSe), the Kesterite $Cu_2ZnSn(S,Se)_4$ (CZTSSe) has a tetragonal crystal structure of Zinc Blende [7]. CZTSSe is a p-type semiconductor with a tunable bandgap between 1.0 and 1.5 eV, with direct transitions and a high absorption coefficient in the range of 10^4–10^5 cm^{-1} [8,9]. For several years, the record efficiency of the CZTSSe device, regardless of the deposition method was 12.6%, as reported in 2014 by IBM [10]. Despite significant efforts in recent years to improve the working characteristics of the CZTSSe solar cells, there are several notable issues that lead to weak performance parameters—deep level defects, a narrow phase stability region, and non-ideal device architecture. There is an approach to the modification of properties of the absorber material by doping or alloying with additional constituents as alkali dopants (Li, Na, K) or isoelectronic substitutions, but there are no reports of real improvements [11]; rather, effective management of the intrinsic defects was suggested to be the way to optimize the optoelectronic properties of the Kesterite-absorber materials.

An extensive and scrupulous analysis of strategies to improve the work characteristics of the Kesterite solar cells was recently presented [12]. In the abovementioned problems with bulk defects, special attention is paid to those in junctions with a buffer and back contact layers. The recently developed new record Kesterite solar cells with 13.6% efficiency is impressive [13]. An epitaxial Kesterite/CdS interface was reconstructed by the low-temperature-induced migration of Zn^{2+} and Cd^{2+}, which were initially disordered by the solution treatment used for deposition of the buffer layer.

Low-temperature technology can enlarge the substrate material base by employing lightweight and flexible carriers [14]. For instance, there is the suggestion that flexible monograin CZTSSe solar cells can be built by combining a high-temperature synthesized absorber material in the structure [15]. In all cases, the optimal way to build thin-film Kesterite solar cells seems to be [12] the initial sintering process of the thin film absorber material, followed by a fast annealing process to avoid components' diffusion.

The aim of this work is to investigate the features of phase composition, films' morphology and a defect distribution in the results of the rapid thermal annealing (RTA) of electrodeposited CZTS thin precursor films. On the basis of the results obtained by XRD and Raman structural analysis, the phase composition is determined, containing the target CZTS and distribution of concomitants. In addition, assessment of the ratio of intensities of specific Raman signals can provide idea for the formation tendencies of some specific defects (as Cu/Zn substitutions) and approaches to the management of structural features by process parameters.

2. Materials and Methods

Substrate layers were electrochemically deposited in potentiostatic conditions using the classic three-electrode cell configuration. Working electrodes were 2 × 1 cm^2-sized tin-oxide-covered soda lime glasses positioned against platinum gauze and a saturated mercury sulfate (MSE) reference electrode (0.6151 V vs. standard hydrogen electrode). Electrolytes for the electrodeposition of Cu-Zn-Sn-S layers were 4 M KCNS aqueous solutions of 0.4 M sodium acetate buffer containing 9 mM of the chloride salts of Cu^+, Zn^{2+}, $SnSO_4$ and $Na_2S_2O_3$ in related ratios. The composition of the as-deposited Cu-Zn-Sn-S layers was copper-poor and zinc-rich with sulfur deficiencies when compared to the stoichiometry. The substrate's influence on the composition of the layers was estimated to be negligible. The process design has been described in detail elsewhere [16]. Reactive annealing was performed in a cold walls' quartz tubular reactor under a flow of 5% H_2S in N_2 at atmospheric pressure. Temperature profile was controlled in dynamic equilibrium by an IR Ulvac-RICO heating system supplied with a PC-driven controller. The system works under continuous cooling provided by the chiller-supplied circulation of working fluid at 10 °C and blowing with 0.6 MPa pressurized air. The heating was achieved by direct IR irradiation on the length of the reactor with xenon lamps. Under these conditions, at every segment, the process temperatures differed from set points with values of less than 0.1%. Below 220 °C, the ramp-down process temperature did not follow strictly the set point values, but it is believed that reactions here are faded. An example work temperature profile is shown in Figure 1.

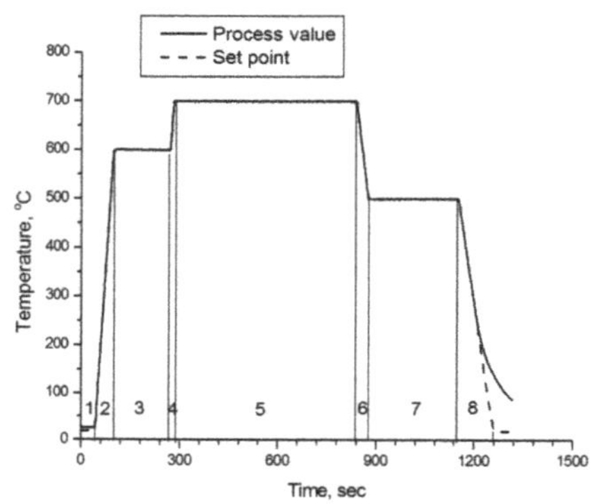

Figure 1. Temperature profile of rapid thermal processing of CZTS films.

The parameters for specific program configuration are summarized in Table 1. After the work cycle, the reactor volume was rinsed by N_2 stream for 5 min.

Table 1. Process parameters of rapid thermal annealing of thin-film CZTS.

Probe	Base	Duration	Duration	Speed Fast	Speed Slow	T °C 750 °C	T °C 650 °C	T °C 600 °C
Sample	1	2	3	4	5	6	7	8
1 Seg	19 °C 10 s	19 °C 10 s	19 °C 10 s	19 °C 10 s	19 °C 10 s	19 °C 10 s	19 °C 10 s	19 °C 10 s
2 Seg	600 °C 1 min	600 °C 1 min	600 °C 1 min	600 °C 30 s	600 °C 3 min	600 °C 1 min	600 °C 1 min	600 °C 1 min
3 Seg	600 °C 3 min	600 °C 3 min	600 °C 3 min	600 °C 3 min	600 °C 3 min	600 °C 3 min	600 °C 3 min	600 °C 3 min
4 Seg	700 °C 10 s	700 °C 10 s	700 °C 10 s	700 °C 5 s	700 °C 30 s	750 °C 10 s	650 °C 5 s	600 °C 5 s
5 Seg	700 °C 15 min	700 °C 10 min	700 °C 5 min	700 °C 15 min	700 °C 15 min	700 °C 15 min	700 °C 15 min	700 °C 15 min
6 Seg	500 °C 40 s	500 °C 40 s	500 °C 40 s	500 °C 20 s	500 °C 2 min	500 °C 45 s	500 °C 15 s	500 °C 15 s
7 Seg	500 °C 5 min	500 °C 5 min	500 °C 5 min	500 °C 5 min	500 °C 5 min	500 °C 5 min	500 °C 5 min	500 °C 5 min
8 Seg	19 °C 2 min	19 °C 2 min	19 °C 2 min	19 °C 1 min	19 °C 10 min	19 °C 2 min	19 °C 2 min	19 °C 2 min

Scanning electron microscopy and energy-dispersive X-ray analysis (EDAX) was performed on a Hitachi TM 1000 unit supplied with an X-ray source (Hitachi, Tokyo, Japan) and detector equipment at an accelerating voltage of 15.0 kV and acquisition time of 90 s. The X-ray diffraction (XRD) analysis was performed by a Rigaku Ultima IV diffractometer (Rigaku, Tokyo, Japan) with Cu-Kα radiation (λ = 1.5418 Å) at 40 kV accelerating voltage. The diffracted beam was scanned in steps by 0.01° for 2 s in an angular range from 10 to 80 degrees 2θ.

Qualitative phase analysis was performed on specialized software PDXL Rigaku's ICDD PDF2-phase research platform (Rigaku, Tokyo, Japan) using the specifications by International Center for Diffraction Data [17] and EDAX data for chemical composition. The suggested faze distribution by XRD is based on the detection of three or more main specific reflections of the phase compared with the actual database cards available at that time [17]. In addition, the broadening of the XRD peaks was used in an analysis of crystallites' size and internal strain. This was more popular than the Willimson–Hall method, which provides an idea of the deviations from the ideal crystalline lattice. The room-temperature (RT) micro-Raman spectra were recorded on a Horiba LabRam 800 high-resolution spectrometer (Horiba Ltd., Kyoto, Japan) equipped with a multichannel detector on backscattering regime. Light source was a red laser with a 633 nm wavelength focused on an at least 10 μm spot diameter, providing a spectral resolution of 0.5 cm^{-1}.

3. Results and Discussion

The reactive annealing is performed in compliance with both the suggested reaction paths and structural features in [18]. To conform with the proposed reaction sequence, the synthesis and structure formation of CZTS begin with the formation of binaries of Cu_2S and SnS_2, followed by their interaction with the ternary Cu_2SnS_3 which, at higher temperatures, reacts with ZnS and completes as Cu_2ZnSnS_4 at temperatures higher than 600 °C. In this sense, the temperature profile configuration is directed to modifications of the area in which the direct formation of the quaternary CZTS takes place.

A detailed analysis of results for rapid thermal annealing consists of a scrupulous phase analysis and assessment of the influence of annealing process parameters on the crystal cell parameters of the target phase—Kesterite (Cu_2ZnSnS_4). In contrast to Stannite, where [19,20] fine differences exist between some closely disposed reflexes of $Cu_2ZnSnSe_4$ and either ZnSe or other binaries, and could become apparent through a more precise angle-resolved analysis, the case for Kesterite is rather more complicated [21] and an additional method such as Raman shift is essential for the correct evaluation of phase distribution.

Figure 2 shows XRD patterns of samples 1, 2 and 3 annealed with different durations (15 min, 10 min and 5 min, respectively) of the Segment 5 at 700 °C, according to Figure 1.

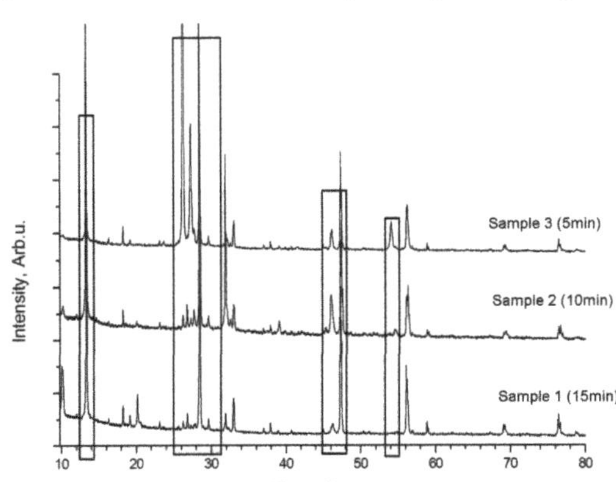

Figure 2. XRD pattern of samples annealed with different durations of Segment 5 (see Figure 1).

The patterns are dominated by characteristic reflexes for the Kesterite phase [1-01-075-4122] at 28.4° (1,1,2), 47.3° (2,2,0) and 56.1° (3,1,2) and its minorities at 16.32°, 18.23°, 23.13°, 29.6°, 32.9°, 36.9°, 37.9°, 40.7°, 44.9°, 56.9°, 58.8°, 67.3°, 69.1°, 76.4° and 78.8° in 2θ scale. Additional appreciable reflexes are detected near 10.2°, 13.3°, 26.2° 54° and 66.1° 2θ and confirm that the system is definitely not a monophase. The formal phase analysis performed through Ref. [17] and presented in Table 2 have specified the content of Sample 1: Kesterite, α-SnS [1:01-083-1758]; Cu_7S_4—[1:00-024-0061] and CuS—[1:03-065-3928].

Table 2. Phase composition according to XRD [17] and Raman shift analyses. •—means registered phase; ◊—unregistered phase; Raman spectrum—−/+ (un)registered phase; ± not sure.

Sample	Process	Cu_2ZnSnS_4 Kesterite-4122	ZnS h	SnS	CuS Covellite	Cu_7S_4 Roxbyite	Cu_2S digenite	ZnS h	ZnS h	ZnS(W) 0688h
1	Base 15 min, 700 °C	•+	•$_{2201}$+	◊	◊	◊	◊−	•$_{2424}$	•$_{2195}$	◊
2	Time—10 min	•+	•$_{2347}$	•+	•	•	◊−	◊	◊	•
3	Time—5 min	•+	◊	•+	•	•	◊−	◊	◊	◊
4	Speed—Fast	•+	•$_{4998}$+	•	◊	◊	◊±	◊	◊	◊
5	Speed—Slow	•+	•$_{6022}$+	◊	◊	◊	◊	◊	◊	◊
6	Temp—750 °C	•+	•$_{2140}$	◊−	◊	◊	•$_{9133/-}$	◊	◊	◊
7	Temp—650 °C	•+	•$_{4989}$	•+	◊	◊	◊±	◊	◊	◊
8	Temp—600 °C	•+	•$_{6009}$	•+	◊	◊	◊	◊	◊	◊
1	Base—700 °C	•+	•$_{2201}$+	◊	◊	◊	◊±	•$_{2424}$	•$_{2195}$	◊

As mentioned above, the main phases, concomitant with the Kesterite dispose reflexes, are quite close to those for the main phase and a more precise analysis could reveal the fine structure of the films. Figure 3 presents part of the pattern of Figure 2 in the vicinity of 13–14° 2θ. The picture presents the fine disposition of reflexes, which allowed for the one for ZnS at 13.35° 2θ (Sample 3) to be distinguished from the other at 13.28° 2θ (Sample 2) belonging to Cu_7S_4.

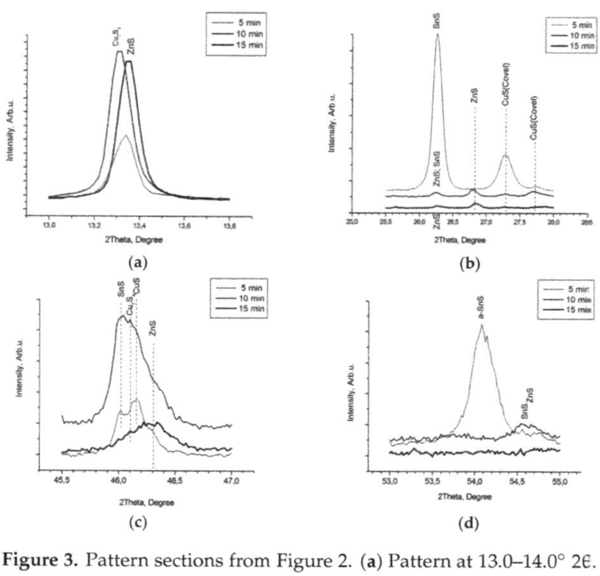

Figure 3. Pattern sections from Figure 2. (**a**) Pattern at 13.0–14.0° 2θ. (**b**) Pattern at 25.5°–28.0° 2θ. (**c**) Disposed pattern between angles 45.5°–47.0°. (**d**) Pattern at 54°.

Figure 3b presents pattern from Figure 2 in the interval 25.5°–28.0° 2θ. In accordance with the performed phase recognition, the reflex at 26.24° for Sample 1 should be assumed to come from SnS only; for Sample 3, the small reflex at the same angle should be assumed for ZnS only, but for Sample 2, the reflex should be consistent with ZnS and SnS.

At 26.82°, 2θ are the disposed reflexes of only Samples 2 and 3, recognized as being from ZnS, whereas, at 27.28° and 27.75° 2θ, single signals can be seen from CuS for Samples 1 and 2. Further, in Figure 3c, patterns between angles 45.5°–47.0° 2θ are shown. Here, the large signal from Sample 3 is due to ZnS. For Samples 1 and 2, the reflex at 46.02° is proven for SnS, whereas those at 46.11° and 46.18° are caused by Cu_7S_4 and CuS, respectively.

Figure 3d shows the field near 54° 2θ in detail. The reflex at 54.05° for Sample 1 comes from α-SnS [1:01-083-1758], whereas, for Sample 2, the reflex at 54.5° is due to SnS [1:01-073-1859] accompanied by Wurtzite [1:00-012-0688].

Figure 4a shows XRD patterns for Sample 4 (fast) and Sample 5 (slow), which, in comparison with Sample 1(base), present an RTA configuration with different rising and decreasing temperature speeds for the dynamic segments No. 2, No. 4, No. 6 and No. 8 (see Figure 1). The right values of rising speed are 3.3 °C/s, 10 °C/s and 20 °C/s, whereas at decreasing steps they are 1.33 °C/s, 5 °C/s and 10 °C/s, respectively.

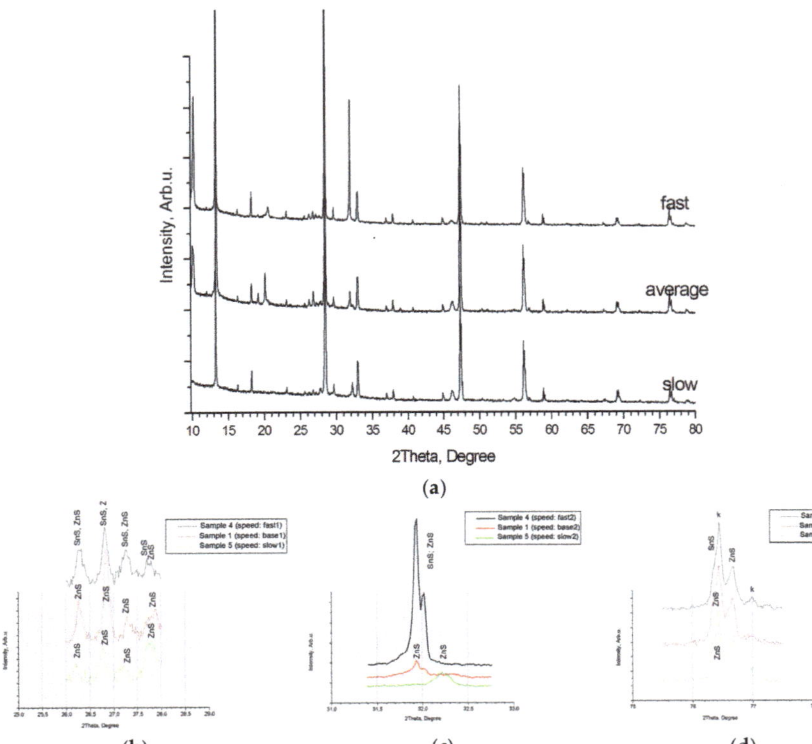

Figure 4. XRD patterns of samples annealed at different speeds (see Figure 1—Segments No.: 2, 4, 6 and 8). (**a**) Superimposed diagram of all speeds. (**b**) Patterns in vicinity of 26°–28° 2θ. (**c**) Patterns in vicinity of 31.5°–32.5°. (**d**) Patterns in vicinity of 76°–77°.

According to Ref. [22], the phase distribution of the samples is summarized in Table 2. Sample 1, annealed at faster temperature increases, contains mainly Kesterite. A pair of SnS phases—orthorhombic [1:01-072-8499] and [1:00-001-0984]—are registered, as well as a Hexagonal ZnS [1:01-074-4998]. For Sample 5, annealed at a slow rate, Kesterite and Hexagonal ZnS [1:01-075-6022] are found. As previously mentioned, the average rate of Sample 1 contained Kesterite and Hexagonal zinc sulphides. A comparative phase analysis shows the Kesterite and zinc sulphides as common structures in the three samples, whereas Sample 4 contains two forms of Herzenbergite [1:01-072-8499 and 00-001-0984]. They have a similar layered orthorhombic structure and differ in their orientation. Obviously, both of them are a result of non-equilibrium interactions in the chain of the formation of CZTS caused by the high-speed changes in the temperature. Apparently, under softer conditions, such as a longer reaction time (previous case) or slower dynamic process parameters, in accordance with Ref. [23], this intermediate further interacts in a related scheme. Figure 4b shows the XRD pattern of the samples in the vicinity from 26° to 28° 2θ. Reflexes near 26.2° are assumed, for all samples, to be from ZnS, but for Sample 4 they appears to be in convolution with a signal from SnS. In the same way, the reflexes at 26.8° and at 27.3° 2θ can be interpreted. Near 27.75° 2θ, the disposition is quite similar but the complex reflex for Sample 4 looks well-divided for both SnS and ZnS. Figure 4c presents the interval 31.5°–32.5° 2θ. Sample 5 disposes an individual reflex at 32.23°, whereas the other Samples, 1 and 4, present a similar twin reflex at 31.94° and 32.02° 2θ. Reference [22] spelt out the twin for Sample 1 as coming from ZnS only, while, for Sample 4, the pair of ZnS and SnS is assumed. This could be explained by the obvious difference in the height (intensity) of the reflexes. Between 76° and 77° 2θ, Figure 4d shows disposed reflexes at 74.4° and 77°, assumed to be Kesterite. ZnS is presented here, with a reflex at 76.68° and a shoulder at the first Kesterite reflex. For Sample 4, a twin shoulder is distinguished, caused by both ZnS and SnS. The Raman spectra presented in Figure 5 are in good agreement with the supposed phase compositions of Samples 4, 5 and 1. In all samples, the Kesterite is well-defined, with the majority of its characteristic vibrations occurring at 338 cm^{-1}, 289 cm^{-1}, 251 cm^{-1}, 98 cm^{-1} and 375 cm^{-1}. For all samples, a shoulder at 352 cm^{-1} is seen, near to the most intensive signal for Kesterite, attributed to ZnS. For Sample 4, a resonant vibration at 164 cm^{-1} is registered, attributed to SnS, in confirmation with the established XRD phase distribution.

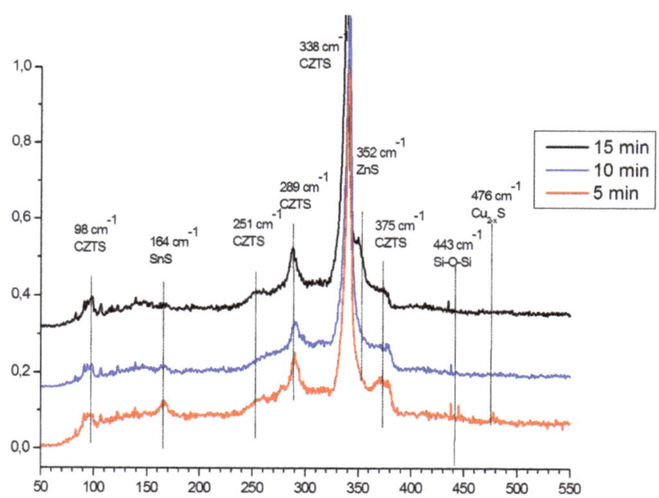

Figure 5. Raman shift—duration.

Figure 5 presents the Raman spectra of the same films. In accordance with [23–25], it can be concluded that the spectra are again dominant for Kesterite signals at 338, 287, 252, 374 and 97 cm^{-1}. After [26,27] the disposition of the shoulder near 350 cm^{-1}, in conjunction with signals at 98 cm^{-1} and possibly near 160 cm^{-1}, the distribution of ZnS can be revealed. Similarly, the shift at 164 cm^{-1} and the shoulder near 311 cm^{-1} confirm the presence of SnS [8,22,25,28,29]. Vibration at 476 cm^{-1} is characteristic of the Cu-S bond in nonstoichiometric unsaturated copper sulphides [27]. This comparative analysis confirms this possible suggested phase distribution, but full identification requires a more detailed view of the XRD patterns.

Figure 6 presents the XRD pattern of Samples 6, 7 and 8, annealed at different temperatures (level of Segment 5, Figure 1) of 600 °C, 650 °C and 750 °C, respectively, which, together with Sample 1 (700 °C for level of Segment 5) are the process set configuration for investigations of the influence of the temperature. The patterns present a well-crystallized Kesterite in all samples, accompanied by different ZnS phases in almost the same structure [1:01-073-6009, 01-074-4989, 01-089-2201, 01-089-2424, 01-089-2195, 01-089-2140].

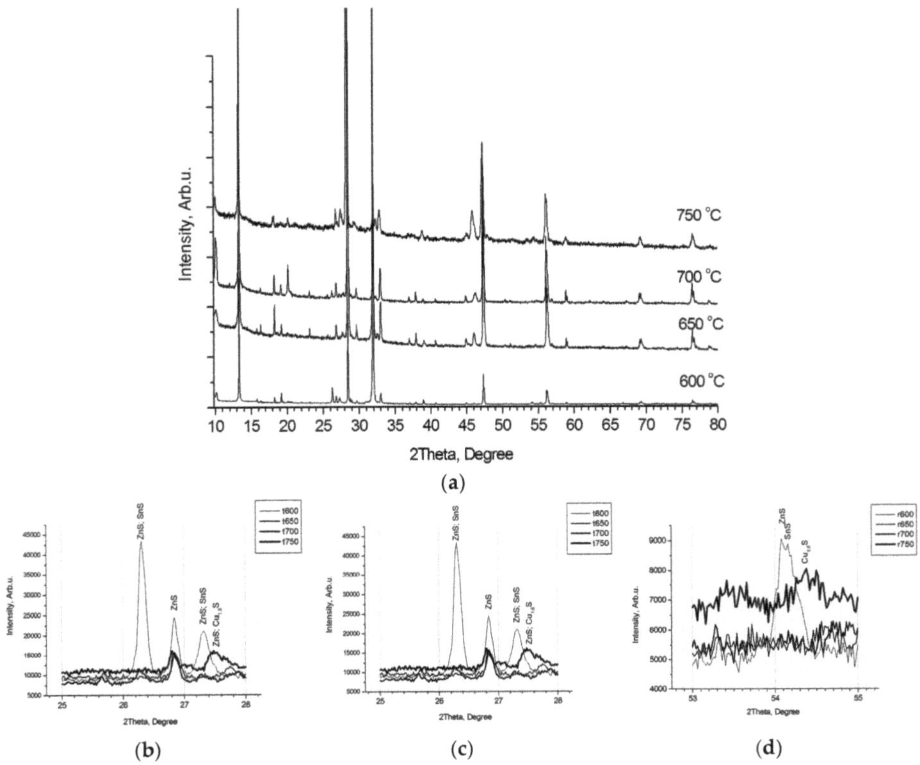

Figure 6. XRD patterns of Segment 5, RT-processed at different temperatures (see Figure 1). (**a**) Superposition at different temperatures. (**b**) Pattern at 26°–28°. (**c**) 45.5°–47°. (**d**) 53°–55°.

For low-temperature Samples 6 and 7, annealed at 600 °C and 650 °C, respectively, hexagonal tin sulphides are registered, while for high-temperature Sample 8 (750 °C), Digenite [1:01-070-9133] is attributed. A detailed XRD analysis in the interval 26°–28° is shown in Figure 6b. Sample 6 disposed three reflexes (at 26.30°, 26.83° and 27.32° 2θ), attributed to ZnS/SnS, ZnS and ZnS/SnS, respectively. The reflex at 26.83° is common for all samples, whereas Sample 8 had a peak at 27.5° 2θ, identified as ZnS that may be $Cu_{1.8}S$. Figure 6c shows the interval of angles between 45.5° and 47°. Samples 5 and 6 (600 and 650 °C) showed reflexes at 46.08°, recognized as being caused by SnS. Between 46.2° and 46.4°, a large reflex is seen for Sample 1, attributed to ZnS. Sample 8 showed a peak between 45.80° and 45.82° 2θ, attributable to ZnS, which may be Digenite, and a shoulder at 46.08°, recognized to be caused by SnS and which may be Digenite. Further, in Figure 6d at 53°–55° 2θ, Sample 5 (600 °C) revealed a twin at 54.09° and 54.16°, whereas for Sample 8 (750 °C), a signal from Digenite was recognized at 54.38°.

Figure 7 presents a set of patterns for Raman shifts in the same samples. Raman spectra showed the presence of Kesterite, ZnS and SnS for Sample 5 and Sample 6, and Kesterite and ZnS for Sample 1 and Sample 8, but there was no confirmation of an Cu-S bond. In sequence, the phase composition features recognized from XRD [22] analysis are confirmed by Raman shift, with the exception of the phase $Cu_{1.8}S$. XRD reflexes were not uniquely defined and the absence of specific Raman resonant vibrations provides reasons to consider this phase as non-existent.

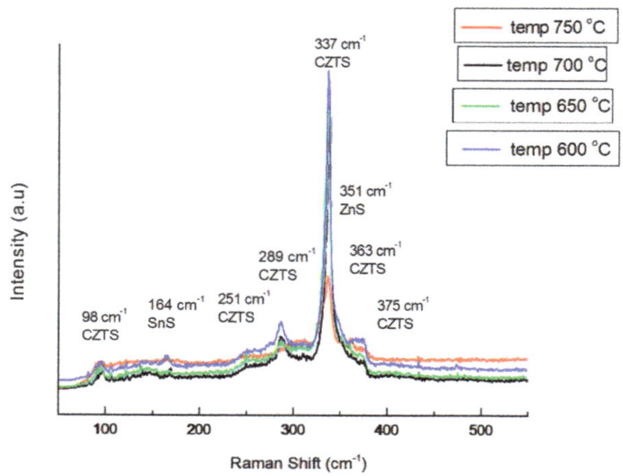

Figure 7. Raman shift for samples annealed at different speeds.

For Sample 2 (10 min), the same phases were recorded, but SnS is in another structure configuration [1:01-073-1859], in addition to ZnS [1:01-089-2347 and 00-012-0688]. Sample 3 (15 min) is common for all other RTA configurations and consists of only Kesterite and Hexagonal ZnS [1:01-089-2201, 01-089-2424 and 01-089-2195].

Figure 8 presents the dependences of crystallite size and lattice strain, calculated by the Williamson–Hall method [17] on process parameters for sets used to investigate the influence of the duration of Segment 5 (Samples 1, 2 and 3), speed of temperature changes in dynamic segments 2, 4, 6 and 8 (Samples 4, 1 and 5), and temperature level of Segment 5 (Samples 6, 7, 1 and 8) at 600, 650, 700 and 750 °C, respectively.

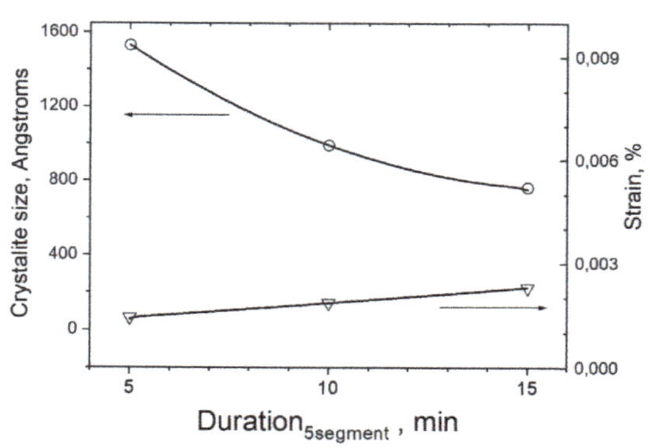

Figure 8. Crystallites size and strain of Kesterite formed by RTA at different durations.

As can be seen, with the increase in duration of Segment 5 from 5 to 15 min, the crystallite size decreases from 1530 Å to near 760 Å, whereas the strain of the cell rises from 0.00143% to 0.00230%. The same trend is seen for the temperature, where the crystallite size drops from 1193 Å for 600 °C to 609 Å for 750 °C, respectively, and the strain rises from 0.00133% to 0.0036%.

Regarding the dependence of crystallite size and strain on the speed of changes in temperature, the general trend is the same, but the curves are not monotonous, as can be seen in Figure 9. The crystallite size is 1521 Å for the fast process and decreases to 1010 Å for the slowest process, while, at average speed, the size is as low as 761 Å. The strain drops from 0.00139% to 0.00026% while, at average speed, the maximum is 0.0023%.

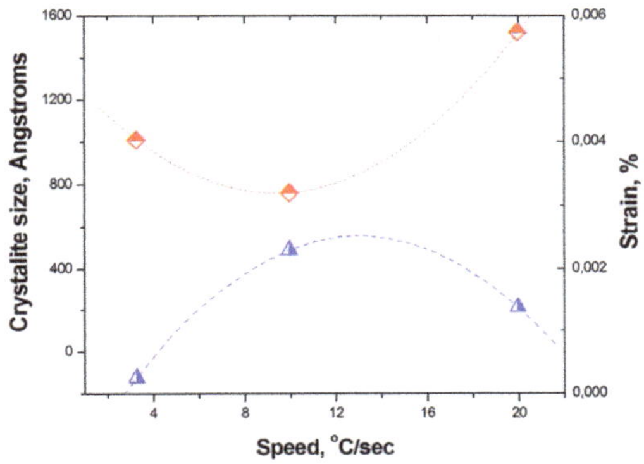

Figure 9. Crystallite size and strain of Kesterite formed by RTA at different speeds.

Regarding the influence of temperature, as shown in Figure 10, in comparison with the data presented in [29], a remarkable difference can be seen. While the cell parameters for 700 °C at [29] are a = 5.6177 Å and c = 11.2232 Å, respectively, in our case they are a = 5.4262 Å and c = 10.8519 Å. This is the reason that a difference in the cell volume is registered at 600 °C from 335 Å^3 [29], whereas cell volume in our case is near 320 Å^3.

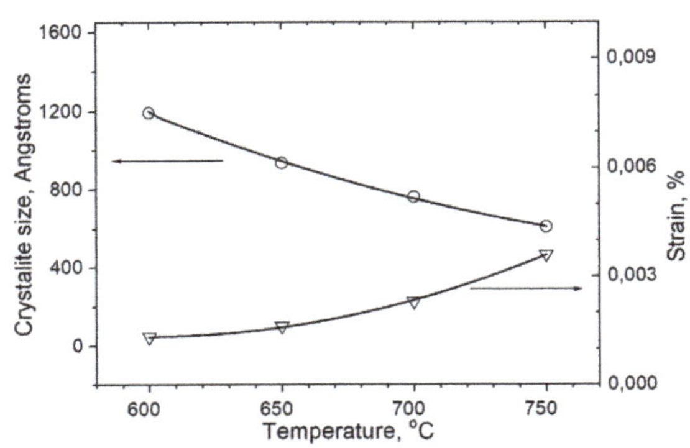

Figure 10. Crystallite size and strain for Kesterite phase formed by RTA at different temperatures.

Generally, it can be concluded that a slower and longer process at higher temperatures for copper-poor, zinc-rich, near-stoichiometric substrates leads to a preferable phase composition of Kesterite and zinc sulphides, whereas the cell strain increases with increasing temperature and an increase in the duration of Segment 5, but decreases when the speed of the process decreases. The results are in agreement with Ref. [8] and Ref. [30], but differ from the newer works in some details [29,31]. For example, the difference in cell parameters [29] at 700 °C is easy to explain when considering the dynamics of the process—in Ref. [29], the process nears equilibrium whereas, in our case, a time-resolved (faster) process is shown, which, in slower stages, coincides with previous results [29–31].

Figure 11 presents SEM micrographs of samples, annealed in different duration.

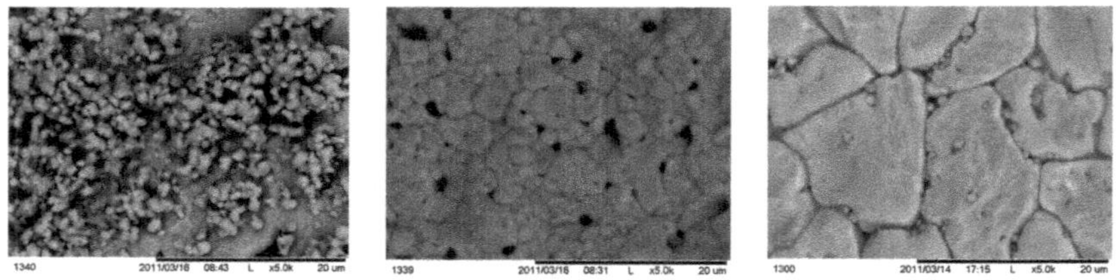

(a) (b) (c)

Figure 11. SEM footprints of samples, annealed at different times (Segment 5, Figure 1): (**a**) 5 min; (**b**) 10 min; (**c**) 15 min.

All scale bars in the figures are equal to 20 μm. For a short annealing time, a small part of the as-deposited layer crystallized on the substrate area, leaving unreacted mass on the top surface. For 10 min (Figure 11b), a pin-holes layer is formed with grain sizes between 1 and 4 μm, whereas at 15 min (Figure 11c), the grains are much larger, with a size 4–5 μm.

Figure 12 presents morphology issues for samples annealed at different speeds. The grain sizes enlarge from 1–2 μm for fast (Figure 12a) annealing, through 4–5 μm for average annealing (Figure 12b), and reaching 8–10 μm for samples annealed using a slow process (Figure 12c).

(a) (b) (c)

Figure 12. SEM fingerprints of samples RTA annealed at different speeds: (**a**) fast process, (**b**) average speed; (**c**) slow process.

Figure 13 shows the morphology footprints for samples processed at different temperatures. Figure 13a presents a CZTS layer with single independent grains with sizes near 1–2 μm. At a temperature of 650 °C, the surface concentration of the grains increased, and pinholes are seen. At 700 °C, the grain sizes reach a maximum of 4–5 μm while, at a higher temperature if 750 °C, the sizes grew smaller with heterogeneous residuals on the grain borders.

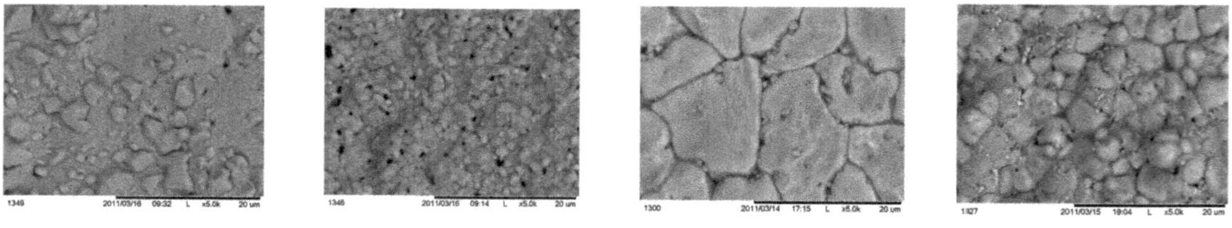

(a) (b) (c) (d)

Figure 13. SEM micrographs for samples annealed at different temperatures: (**a**) 600 °C; (**b**) 650 °C; (**c**) 700 °C and (**d**) 750 °C.

SEM observations a gave slightly different impression of the crystallite size from the calculated values. As shown in Figures 8–10, following XRD analysis, crystallites sizes ranged from 0.05 to 0.15 μm whereas SEM revealed sizes near 1–8 μm.

Here, we can conclude that the tendencies of changes in crystallite size, derived from XRD data obtained using Williamson–Hall method [17], differ by an order of magnitude and, in addition, are the opposite of the tendencies of changes in grain sizes derived from SEM. In fact, the Williamson–Hall method revealed that short-ordered crystal unit Kesterite sizes are quite different from the visible grain structure of the layers.

As was reported elsewhere [32,33], the Raman spectroscopy could be used for characterizing Cu/Zn disorders and to distinguish the origin of the substitution in slow-cooled Kesterite thin films. The assessment of the level of disorder is very important for the device output. Scrupulous comparative analysis with Nuclear Magnetic Resonance (NMR) data revealed the ratio of specific Raman shifts at 287 cm^{-1} and 303 cm^{-1}, $Q = I_{287}/I_{303}$, and additionally introduced a shift between an intensity of 338 cm^{-1} and the sum of intensities at 366 and 374 cm^{-1}, $Q' = I_{338}/(I_{366} + I_{374})$, which is very sensitive to the Cu/Zn disorder. Here, a rough idea of the dependencies in time-resolved RTA processes is presented.

Figure 14a presents the dependencies of the Raman peak ratios, with intensities of $Q = I_{287}/I_{303}$ and $Q' = I_{338}/(I_{366} + I_{374})$, with increases in the duration of the high-temperature annealing step in this process. It can be seen that both Q and Q′ are higher than 1 in all cases, which is characteristic of an ordered material [32] and increase with increases in the duration. The orders-of-magnitude of increases in the Q are quite small—from 1.39 through 1.5, up to 1.86—whereas the Q′ is quite higher and increases by almost twofold, from 2.9 up to 4.7. The same trend is observed regarding the dependence of Q and Q′ on the temperature (level of segment 5), but at the higher level of 750 °C, both Q and Q′ drop sharply. This effect could be interpreted as an increase in the disorder of the Kesterite structure. The increase in the speed of the increase and decrease in temperature influence the Q and Q′ in the opposite way. This effect can be explained by increases in the order of the structure at high speeds in the dynamic stages, bringing the process closer to the case of quenching in the experiments in [33].

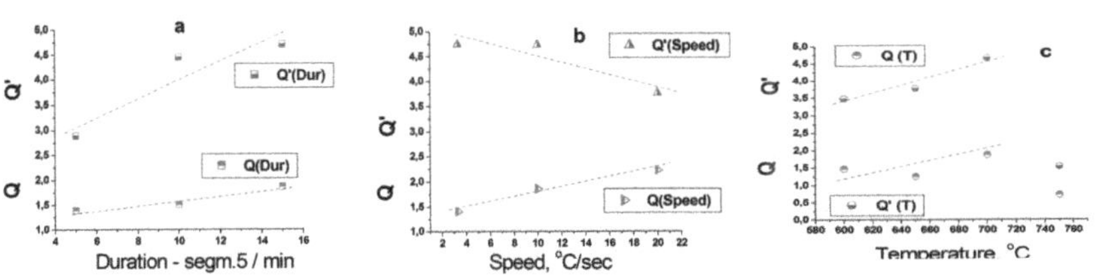

Figure 14. Dependences of Raman peak ratios $Q = I_{287}/I_{303}$ and $Q' = I_{338}/(I_{366} + I_{374})$ on the RTA process parameters: (**a**)—duration of the segment 5; (**b**)—speed of the change in temperature in the dynamic steps—segments 2, 4 and 6; and (**c**)—temperature of segment 4 (see Table 1).

4. Conclusions

A time-resolved annealing process is performed for CZTS thin films. The peculiarities of the phase composition and structure parameters of the formed thin Kesterite layers regarding the dependence of these process parameters (such as time, speed and temperature) are analysed. It is shown that the phase composition of the layers depends on the time and speed of the stage of Kesterite formation. Soon, only unreacted binary residuals, such as copper sulphides and tin sulphides, are left. The dependence is the same for a fast process where the dynamic change in temperature does not allow for binary reaction precursors to react up to the quaternary compound. If the reaction is slow and long enough (at least 15 min in our case), low-temperature binary precursors, according to Schurr et al. [18], react fully and only Kesterite and the high-temperature ZnS are left. The situation is the same regarding the temperature of the CZTS formation process. At low temperatures, unreacted binary copper and tin sulphides are registered, whereas at high temperatures (700–750 °C), well-crystallised Kesterite is formed with ZnS immersions, which is a favourable phase composition regarding the photovoltaic properties of absorber films. However, the time-dependent properties of the Kesterite are elucidated. It is shown that crystallite size decreases with increases in the time needed for the CZTS synthesis stage of the process. The strain in the cell rises in the same way. Generally, the same trend is observed for the process performed at different speeds—for fast processes, the crystallite size is comparatively large and decreases with decreases in the speed as the strain rises. When the temperature rises, the crystallites size decreases, which is in formal contradiction with the results presented by Schurr et al. [18]. However, while the treatments used by Schurr et al. [18] lead to reactions near the equilibrium and it is reasonable to register well-crystallized phases. The time-resolved process requires a shorter time for the equilibrium disposition of the atoms in the structure and formation of the phase that appears in non-equilibrium conditions. As a consequence, the strain in the crystal cell increases. The trends evolved from the RTA features show the better crystallization we can obtain using longer and slower process at comparatively lower temperature limits in the process conditions pointed out in this work. Generally, the order of the structure increases with the increasing duration, speed and temperature of the process, whereas it drops at higher temperatures near 750 °C. The RTA favours the thermal budget of thin absorber films' processing and an appropriate optimization will increase the cost–efficiency ratio.

Author Contributions: Conceptualization, T.R., A.M., M.A. and E.M.; Writing—original draft, M.G.; Writing—review & editing, S.S. All authors have read and agreed to the published version of the manuscript.

Funding: Part of this work was performed on equipment from the Research Infrastructure "Energy Storage and Hydrogen Energetics" (ESHER), included in the National Roadmap for Research Infrastructure 2017–2023", approved by DCM # 354 from 2017 and granted by the Ministry of Education and Science of Republic Bulgaria. The authors are thankful for Estonian Research Council grant PRG1023, the Operational Program "Science and Education for Smart Growth" through the project "MIRACle"-Grant No BG05M2OP001–1.002–0011-C02 and COST Action 21148 ReNEW PV and Bilateral programs between the Bulgarian Academy of Sciences and Estonian Academy of Sciences—2022–2025.

Institutional Review Board Statement: Not applicable.

Informed Consent Statement: Not applicable.

Data Availability Statement: Not applicable.

Conflicts of Interest: The authors declare no conflict of interest.

References

1. Woodhouse, M.; Jones-Albertus, R.; Feldman, D.; Fu, R.; Horowitz, K.; Chung, D.; Jordan, D.; Kurtz, S. *On the Path to SunShot: The Role of Advancements in Solar Photovoltaic Efficiency, Reliability, and Costs*; NREL/TP-6A20-65872; National Renewable Energy Laboratory: Golden, CO, USA, 2016. Available online: http://www.nrel.gov/docs/fy16osti/65872.pdf (accessed on 1 July 2022).
2. Richter, A.; Hermle, M.; Glunz, S.W. Reassessment of the Limiting Eciency for Crystalline Silicon Solar Cells. *IEEE J. Photovolt.* **2013**, *3*, 1184–1191. [CrossRef]
3. Yoshikawa, K.; Hayato, K.; Wataru, Y.; Toru, I.; Katsunori, K.; Kunihiro, N.; Toshihiko, U.; Daisuke, A.; Masanori, K.; Hisashi, U.; et al. Silicon heterojunction solar cell with interdigitated back contacts for a photoconversion efficiency over 26%. *Nat. Energy* **2017**, *2*, 17032. [CrossRef]
4. Nakamura, M.; Yamaguchi, K.; Kimoto, Y.; Yasaki, Y.; Kato, T.; Sugimoto, H. Cd-Free Cu(In, Ga)(Se, S)2 Thin-Film Solar Cell With Record Efficiency of 23.35%. *IEEE J. Photovolt.* **2019**, *9*, 1863–1867. [CrossRef]
5. Biplab, S.R.I.; Ali, M.H.; Moon, M.M.A.; Pervez, M.F.; Rahman, M.F.; Hossain, J. Performance enhancement of CIGS-based solar cells by incorporating an ultrathin BaSi2 BSF layer. *J. Comput. Electron.* **2020**, *19*, 342–352. [CrossRef]
6. Yang, S.C.; Lin, T.Y.; Ochoa, M.; Lai, H.; Kothandaraman, R.; Fu, F.; Carron, R. Efficiency boost of bifacial Cu (In, Ga) Se2 thin-film solar cells for flexible and tandem applications with silver-assisted low-temperature process. *Nat. Energy* **2023**, *8*, 40–51. [CrossRef]
7. Paranthman, M.P.; Wong-Ng, W.; Bhattacharya, R.N. *Semiconductor Materials for Solar Photovoltaic Cells*; Springer: Berlin/Heidelberg, Germany, 2016; ISBN 978-3-319-20331-7.
8. Katagiri, H.; Sasaguchi, N.; Hando, S.; Hoshino, S.; Ohashi, J.; Yokota, T. Preparation and evaluation of Cu_2ZnSnS_4 thin films by sulfurization of E-B evaporated precursors. *Sol. Energy Mater. Sol. Cells* **1997**, *49*, 407–414. [CrossRef]

9. Repins, I.L.; Romero, M.J.; Li, J.V.; Wei, S.-H.; Kuciauskas, D.; Jiang, C.-S.; Beall, C.; DeHart, C.; Mann, J.; Hsu, W.-C.; et al. Kesterite Successes, Ongoing Work, and Challenges: A Perspective From Vacuum Deposition. *IEEE J. Photovolt.* **2013**, *3*, 439–445. [CrossRef]
10. Wang, W.; Winkler, M.T.; Gunawan, O.; Gokmen, T.; Todorov, T.K.; Zhu, Y.; Mitzi, D.B. Device Characteristics of CZTSSe Thin-Film Solar Cells with 12.6% Efficiency. *Adv. Energy Mater.* **2014**, *4*, 1301465. [CrossRef]
11. Romanyuk, Y.E.; Haass, S.G.; Giraldo, S.; Placidi, M.; Tiwari, D.; Fermin, D.J.; Hao, X.; Xin, H.; Schnabel, T.; Kauk-Kuusik, M.; et al. Doping and alloying of kesterites. *J. Phys. Energy* **2019**, *1*, 044004. [CrossRef]
12. Wang, A.; He, M.; Green, M.A.; Sun, K.; Hao, X. A Critical Review on the Progress of Kesterite Solar Cells: Current Strategies and Insights. *Adv. Energy Mater.* **2023**, *13*, 2203046. [CrossRef]
13. Gong, Y.; Zhu, Q.; Li, B.; Wang, S.; Duan, B.; Lou, L.; Xin, H. Elemental de-mixing-induced epitaxial kesterite/CdS interface enabling 13%-efficiency kesterite solar cells. *Nat. Energy* **2022**, *7*, 966–977. [CrossRef]
14. Li, J.; Sun, K.; Yuan, X.; Huang, J.; Green, M.A.; Hao, X. Emergence of flexible kesterite solar cells: Progress and perspectives. *NPJ Flex. Electron.* **2023**, *7*, 16. [CrossRef]
15. Timmo, K.; Altosaar, M.; Pilvet, M.; Mikli, V.; Grossberg, M.; Danilson, M.; Raadik, T.; Kauk-Kuusik, M.; Josepson, R.; Krustok, J. The effect of Ag alloying of $Cu_2(Zn,Cd)SnS_4$ on the monograin powder properties and solar cell performance. *J. Mater. Chem. A* **2019**, *7*, 24281–24291. [CrossRef]
16. Iljina, J.; Zhang, R.; Ganchev, M.; Raadik, T.; Volobujeva, O.; Altosaar, M. Formation of Cu_2ZnSnS_4 absorber layers for solar cells by electrodeposition annealing route. *Thin Solid Films* **2013**, *537*, 85–89. [CrossRef]
17. Williamson, G.K.; Hall, W.H. X-ray line broadening from filed aluminium and wolfram. *Acta Metall.* **1953**, *1*, 22–31. [CrossRef]
18. Schurr, R.; Hölzing, A.; Jost, S.; Hock, R.; Voß, T.; Schulze, J.; Kirbs, A.; Ennaoui, A.; Lux-Steiner, M.; Weber, A.; et al. The crystallisation of Cu_2ZnSnS_4 thin film solar cell absorbers from coelectroplated Cu-Zn-Sn precursors. *Thin Solid Films* **2009**, *517*, 2465–2468. [CrossRef]
19. Salomé, P.M.P.; Fernandez, P.A.; da Cunha, A.F. Morphological and structural characterization of $Cu_2ZnSnSe_4$ thin films grown by selenization of elemental precursor layers. *Thin Solid Films* **2009**, *517*, 2531–2534. [CrossRef]
20. Ganchev, M.; Iljina, J.; Kaupmees, L.; Raadik, T.; Volobujeva, O.; Altosaar, M.; Mere, A.; Raudoja, J.; Mellikov, E. Phase composition of selenized $Cu_2ZnSnSe_4$ thin films determined by X-ray diffraction and Raman spectroscopy. *Thin Solid Films* **2011**, *519*, 7394–7398. [CrossRef]
21. Fernandes, P.A.; Salomé, P.M.P.; da Cunha, A.F. Growth and Raman scattering characterization of Cu_2ZnSnS_4 thin films. *Thin Solid Films* **2009**, *517*, 2519–2523. [CrossRef]
22. Price, L.S.; Parkin, I.P.; Hardy, A.M.E.; Clark, R.J.H.; Hibbert, T.G.; Molloy, K.C. Atmospheric Pressure Chemical Vapor Deposition of Tin Sulfides (SnS, Sn_2S_3, and SnS_2) on Glass. *Chem. Mater.* **1999**, *11*, 1792–1799. [CrossRef]
23. Himmrich, M.; Haeuseler, H. Lattice Vibration-Spectra Spectrochim. *Acta A Mol. Spectrosc.* **1991**, *47*, 933. [CrossRef]
24. Altosaar, M.; Raudoja, J.; Timmo, K.; Danilson, M.; Grossberg, M.; Krustok, J.; Mellikov, E. $Cu_2Zn_{1-x}Cd_xSn(Se_{1-y}S_y)_4$ solid solutions as absorber materials for solar cells. *Phys. Status Solidi A* **2008**, *205*, 167–170. [CrossRef]
25. Grossberg, M.; Krustok, J.; Raudoja, J.; Timmo, K.; Altosaar, M.; Raadik, T. Photoluminescence and Raman study of $Cu_2ZnSn(Se_xS_{1-x})_4$ monograins for photovoltaic applications. *Thin Solid Films* **2010**, *519*, 7403–7406. [CrossRef]
26. Yu, Y.-M.; Hyun, M.-H.; Nam, S.; Lee, D.; O, B.; Lee, K.-S.; Yu, P.Y.; Choi, Y.D. Resonant Raman scattering measurements of strains in ZnS epilayers grown on GaP. *J. Appl. Phys.* **2002**, *91*, 9429–9431. [CrossRef]
27. Fernandes, P.A.; Salome, P.M.P.; da Cunha, A.F. A study of ternary Cu_2SnS_3 and Cu_3SnS_4 thin films prepared by sulfurizing stackedmetal precursors. *J. Phys. D Appl. Phys.* **2010**, *43*, 215403. [CrossRef]
28. Chandrasekhar, H.R.; Humphreys, R.G.; Zwick, U.; Cardona, M. Infrared and Raman spectra of the IV–VI compounds SnS and SnS2. *Phys. Rev. B* **1977**, *15*, 2177–2183. [CrossRef]
29. Schorr, S.; Weber, A.; Honkimäki, V.; Schock, H.-W. In-situ investigation of the kesterite formation from binary and ternary sulphides. *Thin Solid Films* **2009**, *517*, 2461–2464. [CrossRef]
30. Schäfer, W.; Nitsche, R. Tetrahedral quaternary chalcogenides of the type Cu_2 II IV $S_4(Se_4)$. *Mater. Res. Bull.* **1974**, *9*, 645. [CrossRef]
31. Sousa, M.G.; da Cunha, A.F.; Fernandes, P.A.; Teixeira, J.P.; Sousa, R.A.; Leitão, J.P. Effect of rapid thermal processing conditions on the properties of Cu_2ZnSnS_4 thin films and solar cell performance. *Sol. Energy Mater. Sol. Cells* **2014**, *126*, 101–106. [CrossRef]
32. Scragg, J.J.S.; Choubrac, L.; Lafond, A.; Ericson, T.; Platzer-Björkman, C. A low-temperature order-disorder transition in Cu_2ZnSnS_4 thin films. *Appl. Phys. Lett.* **2014**, *104*, 041911. [CrossRef]
33. Paris, M.; Choubrac, L.; Lafond, A.; Guillot-Deudon, C.; Jobic, S. Solid-State NMR and Raman Spectroscopy to Address the Local Structure of Defects and the Tricky Issue of the Cu/Zn Disorder in Cu-Poor, Zn-Rich CZTS, Materials. *Inorg. Chem.* **2014**, *53*, 8646–8653. [CrossRef] [PubMed]

coatings

MDPI

Article

Electrophoretic Deposition and Characterization of Er-Doped Bi_2O_3 Cathode Barrier Coatings on Non-Conductive $Ce_{0.8}Sm_{0.2}O_{1.9}$ Electrolyte Substrates

Elena Kalinina [1,2], Larisa Ermakova [3] and Elena Pikalova [4,5,*]

[1] Laboratory of Complex Electrophysic Investigations, Institute of Electrophysics, Ural Branch of the Russian Academy of Sciences, 620016 Yekaterinburg, Russia; jelen456@yandex.ru

[2] Department of Physical and Inorganic Chemistry, Institute of Natural Sciences and Mathematics, Ural Federal University, 620002 Yekaterinburg, Russia

[3] Laboratory of Chemistry of Compounds of Rare-Earth Elements, Institute of Solid State Chemistry, Ural Branch of the Russian Academy of Sciences, 620990 Yekaterinburg, Russia; larisaer@ihim.uran.ru

[4] Laboratory of Solid Oxide Fuel Cells, Institute of High Temperature Electrochemistry, Ural Branch of the Russian Academy of Sciences, 620137 Yekaterinburg, Russia

[5] Department of Environmental Economics, Institute of Economics and Management, Ural Federal University, 620002 Yekaterinburg, Russia

* Correspondence: e.pikalova@list.ru; Tel.: +7-343-362-31-94

Abstract: In this study, the formation of thin-film barrier coatings based on a highly conductive $Bi_{1.60}Er_{0.4}O_3$ (EDB) solid electrolyte on supporting $Ce_{0.8}Sm_{0.2}O_{1.9}$ (SDC) electrolyte substrates was implemented for the first time using electrophoretic deposition (EPD). The electrokinetic properties of EDB-based suspensions in a non-aqueous dispersion medium of isopropanol modified with small additions of polyethyleneimine (PEI, 0.26 g/L) and acetylacetone (0.15 g/L), as well as in a mixed isopropanol/acetylacetone (70/30 vol.%) medium, were studied. The dependences of the thickness of the EDB coatings on voltage and deposition time were obtained using deposition on a model Ni foil electrode. Preliminary synthesis of a conductive polypyrrole (PPy) polymer film was used to create surface conductivity on non-conductive SDC substrates. The efficiency of using a modified dispersion medium based on isopropanol to obtain a continuous EDB coating 12 μm thick, sintered at a temperature of 850 °C for 5 h, is shown. The microstructure and morphology of the surface of the EDB coating were studied. A Pt/SDC/EDB/Pt cell was used to characterize the coating's conductivity. The EPD method is shown to be promising for the formation of barrier coatings based on doped bismuth oxide. The developed method can be used for creating cathode barrier layers in SOFC technology.

Keywords: electrophoretic deposition; solid oxide fuel cell; thin-film electrolyte coating; MIEC electrolyte; barrier layer; doped Bi_2O_3

Citation: Kalinina, E.; Ermakova, L.; Pikalova, E. Electrophoretic Deposition and Characterization of Er-Doped Bi_2O_3 Cathode Barrier Coatings on Non-Conductive $Ce_{0.8}Sm_{0.2}O_{1.9}$ Electrolyte Substrates. *Coatings* **2023**, *13*, 1053. https://doi.org/10.3390/coatings13061053

Academic Editor: Christos Argirusis

Received: 4 May 2023
Revised: 25 May 2023
Accepted: 4 June 2023
Published: 6 June 2023

1. Introduction

The major challenge for the commercialization of solid oxide fuel cells (SOFCs) is to reduce their operating temperature to an intermediate temperature (IT) range of 600–750 °C while maintaining the cell performance as comparable with those of high-temperature devices [1]. An increase in the ohmic resistance of an electrolyte membrane with a decrease in operating temperatures is the main problem for achieving the set goals of maintaining a high performance for IT-SOFCs. Alternative electrolytes having higher ionic conductivity than conventional stabilized zirconia have therefore become increasingly attractive for use in IT-SOFCs [2,3]. To date, most studies on electrolytes possessing satisfactory oxygen ion conductivity at decreased temperatures have been focused on oxide materials with cubic fluorite structures, such as doped CeO_2 [4,5] and Bi_2O_3 [6,7]. Nevertheless, despite high ionic conductivity, which is very attractive for designing IT-SOFCs, these electrolytes, along with having advantages over traditional electrolytes based on zirconium dioxide, have a number of disadvantages.

CeO_2-based materials are thermodynamically stable over a wide range of temperatures, in the presence of water vapor, hydrocarbons, and hazardous gases, and are chemically compatible with a wide range of oxide electrodes. One of the major problems of using CeO_2 electrolytes as SOFCs' membranes is that at low partial pressures of oxygen, partial reduction of cerium $Ce^{4+} \rightarrow Ce^{3+}$ occurs, which leads to the appearance of electronic conductivity in the material and can cause an internal short circuit in the SOFC, resulting in reductions in open-circuit voltage (OCV) and cell power, as well as fuel utilization efficiency [5]. As a solution to the problem of internal circuiting in CeO_2 electrolytes, the formation of blocking layers on the anode side based on yttria-stabilized zirconia or barium cerate-zirconate is used [8]. However, due to their lower ionic conductivity compared with electrolytes based on cerium dioxide, the introduction of such blocking layers may cause a decrease in the overall conductivity of the electrolyte membrane [9]. In this regard, the most promising technological decision is to use a more conductive electrolyte for the blocking layer compared with CeO_2, such as stabilized Bi_2O_3 electrolytes [10,11].

The use of bismuth oxide as a solid electrolyte is limited by its thermodynamic instability under reducing conditions—namely, the oxide is reduced to metallic bismuth at $pO_2 < 10^{-13}$ atm [12]. Improving the thermodynamic stability of Bi_2O_3 is achieved by doping it, for example, with Er or Y, which also makes it possible to increase its ionic conductivity [13]. The formation of blocking layers of doped bismuth oxide on the cathode side of the doped ceria electrolyte is aimed at reducing the internal leakage current in the cell; on the other hand, the main ceria electrolyte prevents the reduction of bismuth oxide by eliminating its contact with the reducing atmosphere. Studies on the effect of a doped Bi_2O_3-based blocking layer on the performance of ceria electrolyte-supported SOFCs showed the possibility of improving the maximum power density (MPD); however, the growth in OCV was less significant than expected and achieved at a blocking layer thickness of no less than 30 μm [14,15]. The low effect on the OCV value can be explained by an insufficient density of Bi-containing blocking layers, which is strongly influenced by the deposition method [16]. The highest OCV for the cell with a Sm-doped (SDC) 1 mm thick electrolyte in the range of 1.006–0.900 V at 500–800 °C was reached by using a $Bi_{1.6}Er_{0.4}O_3$ cathode barrier layer (30 μm) obtained using dip coating [16]. Among other methods used for the deposition of doped Bi_2O_3 coatings are screen-printing [15,17,18], magnetron sputtering [19], and pulsed laser deposition (PLD) [20].

The method of electrophoretic deposition (EPD) is promising for the formation of various barrier layers, since this method ensures rapid formation of coatings of complex chemical compositions on substrates of different shapes (flat or tubular) with a reproducible coating thickness. The advantage of the EPD method is the simplicity of its technological implementation and scalability [21–23]. The application of the EPD method is associated with the preparation of suspensions of powder materials of various morphologies and dispersions. The movement and deposition of powder particles in suspension on the substrate occurs under the influence of an external electric field, which is accompanied by the formation of a coating, which is subsequently subjected to drying and high-temperature sintering. To the best of our knowledge, there are a few published studies on the EPD of Bi_2O_3 [24,25] and $BiVO_4$ [26,27] coatings for photo-electrochemical cells and photocatalytic applications. However, the preparation of stable suspensions based on doped Bi_2O_3 with following electrophoretic deposition on non-conductive dense electrolyte substrates is not a covered topic. In the present work, we have for the first time carried out the formation of a coating based on erbium-doped bismuth oxide, $Bi_{1.6}Er_{0.4}O_3$ (EDB), on an SDC solid electrolyte using the EPD method. The choice of the EDB electrolyte was based on that fact that EDB is stable under the cathode conditions of SOFCs, as this material has been used not only as a cathode interlayer for ceria- or zirconia-based materials [17], but also in a nanocomposite electrolyte in a mixture with a YSZ electrolyte [28], and as a component of composite cathodes for IT-SOFCs [29–32]. The stability of the EDB-containing structures was demonstrated in durability tests from 200 to 580 h. We studied the electrokinetic properties of EDB suspensions in a non-aqueous dispersion medium, the features of the formation and sintering of EDB coatings and their morphology, and the characteristics of a Pt/SDC/EDB/Pt single cell in order to determine the effect of the EDB blocking layer on the OCV values.

2. Materials and Methods

2.1. Synthesis and Characterization of the Electrolytes

Powders of $Bi_{1.60}Er_{0.4}O_3$ (EDB) were synthesized using solution combustion synthesis (SCS). Erbium oxide (Er_2O_3) (99.90% purity, Sigma-Aldrich, St. Louis, MI, USA), bismuth nitrate [$Bi(NO_3)_3$*$5H_2O$] (98.00% purity, Riedel Chemicals, Seelze, Germany), glycine [$C_2H_5NO_2$] (AMK Ltd., Russia), and citric acid [$C_6H_8O_7$] (>98% purity, Weifang Ensign Industry Co. Ltd., Weifang, China) were used as the starting raw materials. An aqueous solution of metal nitrates in distilled water was mixed in stoichiometric ratio. Appropriate amounts of glycine and citric acid (as fuel and complexing agents) were then added to the mixed nitrate solution. The nitrates acted as oxidizer in the mixed solution. The molar ratio of fuel to nitrate was set at 1.4:1 (fuel enrichment range) to ensure complete complexation of Bi and for smooth and controlled combustion of the mixed solution of bismuth and erbium nitrates. In addition, increasing the amount of fuel above stoichiometric (φ = 1) leads to an increase in the amount of gas-phase product, which is an important factor in controlling the product-specific surface area [33]. The ratio of glycine and citric acid was 1:1. The mixed solution was heated on a hotplate under stirring to form a gel, which was further heated until combustion occurred. The foam-like powder obtained after combustion was crushed and calcined at 600 °C for 5 h to remove residual organics. After that, the powder was calcined at 700 °C for 5 h for subsequent phase analysis.

The $Ce_{0.8}Sm_{0.2}O_{1.9}$ (SDC) powder for the supporting electrolyte substrates was fabricated with a solid-state reaction method using CeO_2 (99.9% wt) and Sm_2O_3 (99.9% wt) as the starting reagents. The reagents were mixed in a PM 100 planetary mill (Retsch, St. Petersburg, Russia) in ethyl alcohol medium using plastic drums and stainless-steel grinding bodies, dried, and then calcined in alumina crucibles at 950 °C (10 h) and 1050 °C (10 h) with intermediate ball-milling for 1 h. After the final calcination step, the powder was ball-milled for 1 h, dried, and dry-pressed into disks at 300 MPa. The disks were then sintered at 1600 °C for 3 h. To perform XRD characterization, the SDC sintered disk was crushed.

The XRD characterization of the obtained materials was performed using an XRD-7000 diffractometer (Shimadzu, Kyoto, Japan) in CuKα1 radiation in the $25° \leq 2\theta \leq 80°$ angle range with a scanning step of $\Delta(2\theta) = 0.02°$ and a fixed time of 5 s at each point. The parameters of the crystal structure of the obtained materials were refined using FullProf Suite software [34]. The specific surface area of the EDB powder used for the suspension preparation was determined using a SORBI N 4.1 instrument (Meta, Novosibirsk, Russia). The morphology of the EDB powder was investigated using a JSM-6390 LA scanning electron microscope (JEOL, Tokyo, Japan).

2.2. Preparation of the Suspensions Based on the Electrolyte Materials and Their Characterization

Suspensions based on EDB powder with a concentration of 10 g/L were prepared in a mixed dispersion medium of isopropanol/acetylacetone (70/30 vol.%), as well as in an isopropanol medium with the addition of polyethyleneimine (PEI, 0.26 g/L) and acetylacetone (0.15 g/L). EDB suspensions were sonicated using an ultrasonic bath UZV-13/150-TH (Reltec, Yekaterinburg, Russia) at a generator power of 210 W and an operating frequency of 22 kHz for 125 min at room temperature. The electrokinetic zeta potential and pH in the as-prepared suspensions were measured with the electroacoustic method using a DT-300 analyzer (Dispersion Technology, Bedford Hills, NY, USA).

2.3. Electrophoretic Deposition of the Barrier Electrolyte Layers on the SDC Substrates and Their Characterization

Electrophoretic deposition was performed using a specialized computerized experimental setup (Institute of Electrophysics, UB RAS, Yekaterinburg, Russia) with vertically arranged electrodes. The supporting SDC electrolyte disks were polished using a diamond disk to 550 μm in thickness and 13 mm in diameter. The substrates were cleaned in the ultrasonic bath for 10 min and calcined at 900 °C for 1 h to eliminate surface contamination. A conductive polymer film of polypyrrole (PPy) was synthesized on the surface of the SDC substrate using chemical polymerization of pyrrole in an aqueous solution with ammonium persulfate serving as an oxidizing agent (98%, 0.03 M), sodium salt of p-toluenesulfonic acid serving as a dopant (97.5%, 0.03 M), and pyrrole monomer (98%, 0.03 M) [35]. To perform the EPD process, the SDC substrate with the deposited PPy conductive polymer film was placed on the EPD cell cathode electrode at a distance of 10 mm from the stainless-steel counter electrode, which was 12 mm in diameter. The deposition was performed in the constant voltage mode, and current strength during the deposition process was controlled using an Intelligent Digital Multimeter UNI-T UT71E (Uni-Trend Technology, Guangdong, China). The morphology of the surface of the deposited layers was examined using an ST-VS-520 optical microscope (Yekaterinburg, Russia). The microstructure and element composition of the deposited films were studied by means of a JSM-6390 LA scanning electron microscope (JEOL, Tokyo, Japan) equipped with a system of energy-dispersion X-ray microanalysis (EDX).

2.4. Single Cell Fabrication and Electrochemical Characterization

To study the effect of the EDB blocking layer on the OCV values, a Pt/SDC/EDB/Pt single cell was fabricated. Platinum electrodes with an effective area of approximately 0.3 cm^2 were paint-brushed symmetrically on both sides of the SDC electrolyte disks with the deposited and sintered EDB barrier coating. The electrodes were sintered at 900 °C for 2 h. To reduce polarization resistance of the electrodes, they were activated through impregnation with an ethyl alcohol solution of $Pr(NO_3)_3$ (99.9% wt) (cathode side) and water solution of $Ce(NO_3)_3$ (99.9% wt) (anode side). Nitrate decomposition was performed at 600 °C for 2 h, with a rate of heating/cooling of 100 °C/h.

The measuring cell consisted of the YSZ tube with deposited Pt electrodes serving as electrochemical pump and sensor. The Pt/SDC/EDB/Pt cell was fixed on the top of the measuring YSZ cell, with the Pt anode activated with Ce inside and the Pr cathode activated with Pr outside the YSZ tube. The sample was first fixed using Aremco Ceramabond™ 571 (Aremco Products Inc., Valley Cottage, NY, USA) and then adhered tightly to the tube using a high-temperature sealant through heating up to 930 °C, with a holding time of 10 min, and a heating/cooling rate of 150 °/h. The electrochemical study was performed using a B2901A precision source/measure unit (Keysight, Santa Rosa, CA, USA). The spectra under the OCV conditions were collected using a Parstat 3000A potentiostat/galvanostat (Ametek Scientific Instruments, Newark, NJ, USA) at 30 mV in a frequency range of 2 MHz–0.1 Hz, 30 points per decade. Z-View2 software (version 3.5i, 2021, Scribner Associates, Southern Pines, NC, USA) was used to fit the obtained spectra.

To start the measurements, the cell was heated up to 600 °C and then the oxidizing atmosphere in the anode channel (inside the tube) was gradually replaced with argon and then with humidified hydrogen (3% H_2O, 5 L/h flow rate). The air was pumped to the cathode side at a rate of 5.7 L/h. After the establishment of the necessary atmosphere in the anode channel, the EMF measurements using the electrochemical sensor placed on the YSZ-based measuring cell at 600 °C were equal to 1120 mV, which is characteristic for the YSZ electrolyte [36]. The measurements were carried out at 600–850 °C by sequentially recording the OCV values on the Pt/SDC/EDB/Pt cell, the cell impedance spectrum under OCV conditions, and the cell volt-ampere characteristics. The polarization and ohmic (IR) resistance of the cell were measured using the current interruption method with an RTM3004 oscillograph (Rohde & Schwarz, Munich, Germany).

3. Results and Discussion

3.1. Characterization of the Electrolyte Powder Materials

The results of the XRD characterization of the EDB and SDC electrolyte materials are represented in Figure 1. The materials were single-phase and exhibited a cubic-type structure (Table 1), with the unit cell parameters being close to those presented in the literature [5,37,38].

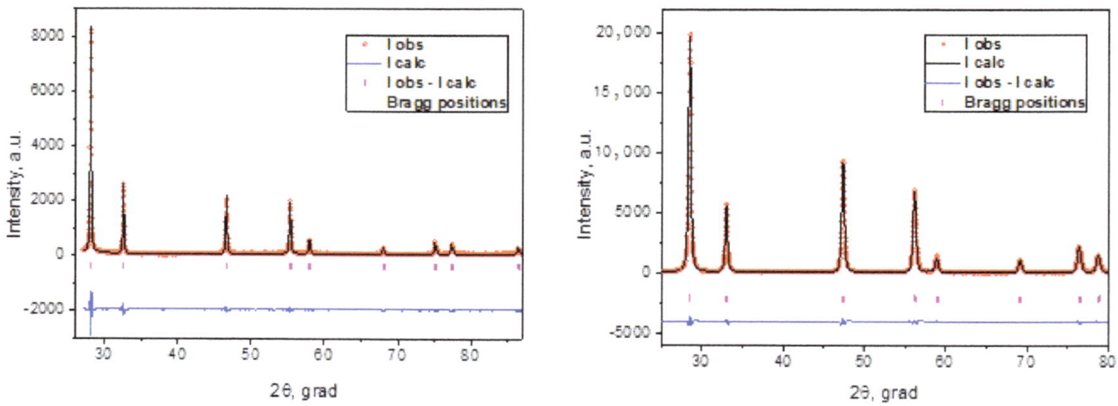

(**a**) (**b**)

Figure 1. Observed (points) and calculated (lines) XRD patterns for $Bi_{1.60}Er_{0.4}O_3$ (**a**) and $Ce_{0.8}Sm_{0.2}O_{1.9}$ (**b**) powders.

Table 1. Structural characteristics of the EDB and SDC electrolyte materials.

Electrolyte Material	Crystal Lattice Type, Space Group	Lattice Parameters, Å
EDB	cubic, Fm-3m	a = 5.4960(1)
SDC	cubic, Fm-3m	a = 5.4324(3)

The powder after synthesis was characterized by the formation of large agglomerates up to 10 μm in size, consisting of irregularly shaped submicron particles (Figure 2a). EDX analysis revealed the uniform distribution of elements in the particles (Figure 2b,c). Along with the presence of agglomerates, the EDB powder was also characterized by the presence of individual particles with a size of 0.1 μm or greater. The morphological inhomogeneity of the powder can lead to an uneven distribution of electrolyte powder particles over the substrate surface and subsequent multidirectional sintering in different parts of the surface during the formation of the EDB electrolyte layer. The crystallite size of the powder calcined at 700 °C, as determined by the Scherrer formula, and was about 175 nm. The powder was additionally ball-milled for 1 h in an ethanol medium. After that, the specific surface area of the EDB powder reached 3 m^2/g with an average particle size of 222 nm, calculated according to [39].

(c)

Figure 2. SEM image of the EDB electrolyte powder (T_{calc} = 700 °C) (SEM image): (**a**) agglomerates, ×6000; (**b**) integral map of elements' distribution; (**c**) and individual element's maps.

3.2. Preparation of Suspensions for EPD and Study of Kinetic Properties

The electrokinetic properties of suspensions of the EDB powder (10 g/L) in an isopropanol/acetylacetone (70/30 vol.%) dispersion medium were studied after ultrasonic treatment (UST) for 5–125 min; the results are presented in Table 2. A slight increase in the zeta potential of up to +20 mV during ultrasonic treatment for 125 min was noted, while the pH shifted from 6.8 to 6.2, which characterizes a change in the ionic composition of the suspension due to improved solvation of the particles by the dispersion medium. In the course of the experiments, another variant of the composition of the dispersion medium for the preparation of the EDB suspension was used: isopropanol with the addition of polyethyleneimine (PEI, 0.26 g/L) and acetylacetone (0.15 g/L). The EDB suspension preparation was carried out in the following sequence. First, the EDB powder was introduced into an isopropanol medium with the addition of PEI (0.26 g/L), and UST was performed for 125 min. After that, the measured zeta potential value was +10 mV at pH = 11.0. However, in a suspension of EDB in an isopropanol medium with the addition of PEI, the deposition process did not occur. In the second step of the preparation, the resulting suspension was further modified by the addition of a small amount of acetylacetone (0.15 g/L). The amount of acetylacetone added as a dispersant was calculated on the specific surface of the EDB powder, namely in the amount of 5 mg/m^2. The principles of introducing acetylacetone as a dispersant were described in our recent study [40]. After the addition of acetylacetone (0.15 g/L) to the suspension of EDB in isopropanol with PEI and UST for 125 min, the zeta potential and pH values were +9 mV and 10.5, respectively. Despite the addition of acetylacetone having no significant effect on the values of the zeta potential and pH, the modification of the EDB suspension with the addition of the acetylacetone dispersant made it possible to carry out the EPD process.

Table 2. Electrokinetic properties of the EDB suspensions (10 g/L) in various dispersion media: isopropanol/acetylacetone (70/30 vol.%); isopropanol-PEI; isopropanol-PEI-HAcAc.

Suspension	UT, min	Zeta Potential, mV (pH)
EDB isopropanol/ acetylacetone 70/30 vol.%	5	+16 (6.8)
	25	+19 (6.6)
	125	+20 (6.2)
EDB isopropanol-PEI	125	+10 (11.0)
EDB isopropanol-PEI-HAcAc	125	+9 (10.5)

3.3. EPD from the EDB Suspensions on a Model Ni Foil Substrate

The selection of EPD modes was carried out on the basis of preliminary experiments on the deposition of the EDB powder from the prepared suspension in the isopropanol/acetylacetone 70/30 vol.% medium onto a model electrode (Ni foil). The dependences of the EDB coating thickness on the voltage at a fixed deposition time (1 min), as well as at a constant voltage of 80 V and various deposition times, are shown in Figure 3. A steady increase in the thickness of the EDB coating was observed in the voltage range of 50–80 V (Figure 3a). The dependence of the coating thickness on time (Figure 3b) was close to linear.

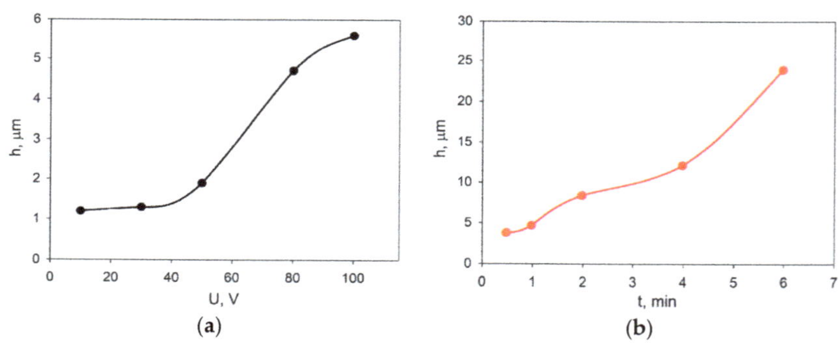

Figure 3. Dependences of the thickness of the EDB coating obtained from suspension in the isopropanol/acetylacetone 70/30 vol.% medium on the Ni foil model electrode: (**a**) at a constant deposition time of 1 min and various voltages; (**b**) at a fixed voltage of 80 V and various deposition times.

A continuous EDB coating on the model substrate was formed at a voltage of 80–100 V (Figure 4a), while at lower voltages an inhomogeneous network coating appeared (Figure 4b). According to the obtained results, for the following deposition on the SDC substrate an EPD mode of U = 80 V and t = 1 min was chosen to obtain a continuous EDB coating with a thickness of 5 μm. The deposition current was ~0.1 mA.

The same scheme was used to study the effect of EPD modes on the coating thickness on a model electrode (Ni foil) deposited from the EDB suspension in an isopropanol medium modified with PEI and acetylacetone. The results are shown in Figure 5.

At a voltage of more than 80 V, the growth of the EDB coating thickness was accelerated (Figure 5a); an increase in the deposition time to more than 4 min led to a slowdown in the growth of the coating thickness (Figure 5b). It was established that to obtain a continuous EDB coating, the deposition must be carried out at a voltage of at least 80 V, and to obtain a thickness of 5 μm, the deposition time must be ~2 min (Figure 6a). Lower voltages and deposition times (e.g., 30 V and 1 min) resulted in a non-uniform coating structure (Figure 6b). Based on the results obtained, for the following experiments a deposition mode of U = 80 V and t = 2 min was chosen to obtain a 5 μm thick EDB coating from a suspension in isopropanol with PEI and acetylacetone additives. The deposition current was 0.1 mA.

Figure 4. Optical surface images of the dried EDB coatings obtained from a suspension in an isopropanol/acetylacetone 70/30 vol.% medium on a Ni foil model electrode: (**a**) at a deposition time of 1 min and a voltage of 80 V; (**b**) at a deposition time of 1 min and a voltage of 30 V.

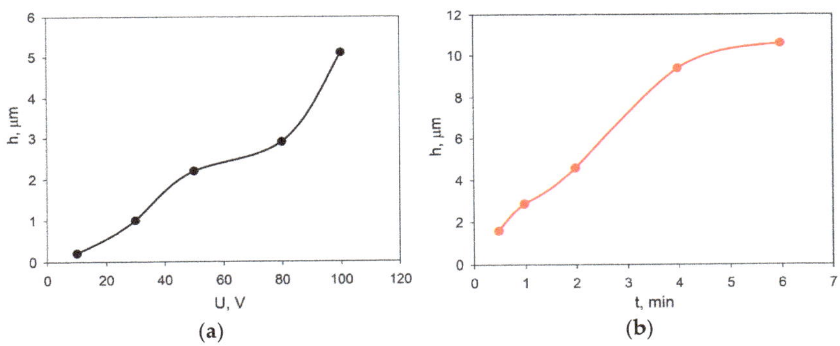

Figure 5. Dependences of the EDB coating thickness, obtained from a suspension in isopropanol with the addition of PEI and acetylacetone, on a Ni foil model electrode: (**a**) at a constant deposition time of 1 min and various voltages; (**b**) at a fixed voltage of 80 V and various deposition times.

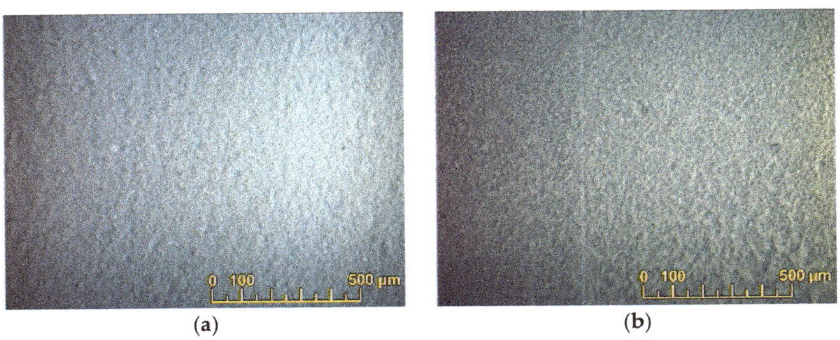

Figure 6. The surfaces of dried EDB coatings obtained from a suspension in isopropanol with the addition of PEI and acetylacetone, on a Ni foil model electrode: (**a**) U= 80 V and t = 2 min; (**b**) U = 30 V and t = 1 min.

3.4. Formation of EDB Coatings on the Dense SDC Electrolyte Substrates

The implementation of the EPD method is related to the need to use a conductive substrate. Using EPD on non-conductive dense (solid-state electrolyte membranes) or low-porosity (NiO-based cermet anodes) substrates is possible by creating a conductive sublayer on their surface. Creating surface conductivity can be achieved through the deposition of conductive graphite layers [41–43], through metallizing the substrate surface with silver or platinum [44], or through the deposition of a polypyrrole (PPy) coating [45]. The synthesis of PPy was chosen in this study. It is known that PPy films exhibit conductivity sufficient for EPD (~500 S/m) and a small thickness, which ensures the integrity of the electrolyte coating when burning PPy during sintering [35,46,47]. In this work, a conductive PPy film was synthesized on the surface of the dense SDC substrate through the chemical polymerization of a pyrrole monomer using the reagents described in Section 2. The reagents were intensively mixed at 0 °C to obtain a homogeneous reaction mixture, into which the SDC substrate was immersed immediately after the start of synthesis. Chemical polymerization of PPy occurs according to the following reaction:

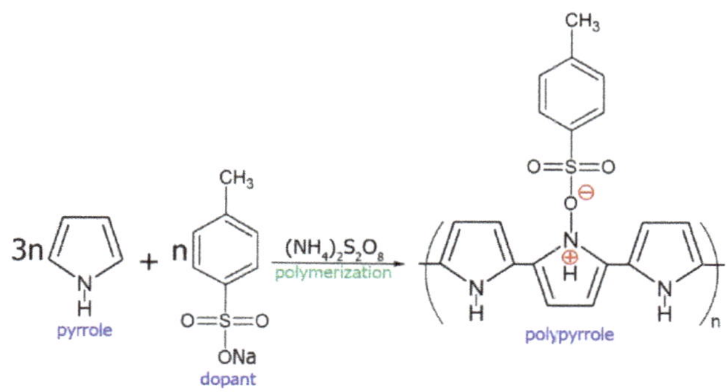

The thickness of the resulting PPy sublayer was approximately 0.5 μm. On the SDC substrate with a PPy sublayer, the EDB barrier coating was deposited from the isopropanol/acetylacetone-based suspension in the constant voltage mode of U = 80 V and t = 1 min (Sample EDB-1). The EDB layer was dried at room temperature in a Petri dish. The thickness of the dried coating was 6 μm. The coating was sintered at a temperature of 850 °C for 5 h in a closed crucible with the EDB powder being poured around the substrate. The EDB coating was uniform and crack-free in the central zone of the substrate (Figure 7a), but breaks in the coating were observed along the edges of the SDC substrate. An attempt was made to close the formed defects in the coating by applying the second EDB layer with EDB. The deposition was carried out according to a similar scheme: a PPy layer was synthesized on the surface of the sintered EDB layer and the second EDB layer was deposited (U = 80 V and t = 1 min), followed by being dried and sintered at a temperature of 850 °C for 5 h. The total thickness of the EDB coating after two deposition–sintering cycles was 11 μm. However, as a result, a net of breaks was observed on the surface of the sintered coating (Figure 7b).

The formation of cracks and breaks in the EDB coating of the EDB-1 sample may be associated with the effect of shrinkage of the EDB electrolyte material. However, individual fragments of the EDB electrolyte layer had a sintered structure, which allowed us to conclude that the selected sintering temperature of 850 °C is sufficient. To reduce the degree of shrinkage and evaporation of the EDB electrolyte material during sintering, we carried out additional annealing of the initial EDB powder at a temperature of 650 °C for 1 h. However, it was found that the annealed powder was not suitable for the preparation of a stable suspension in the isopropanol/acetylacetone 70/30 medium vol.%, and subsequent deposition from this suspension did not occur. Based on the results obtained, the composition of the dispersion medium was changed; namely, the powder was dispersed in an isopropanol medium with PEI and acetylacetone additives, as described in Section 3.2. The following formation of the EDB barrier layer was carried out on an SDC substrate with a PPy sublayer with two deposition–sintering cycles (EDB-2 sample). In the first deposition cycle, a 5 μm thick EDB barrier layer was obtained in the EPD mode at a constant voltage of 80 V for 2 min, after which it was dried and sintered at a temperature of 850 °C for 5 h. The surface SEM image of the resulting EDB coating is shown in Figure 8. The EDB was not continuous; however, no breaks in the material or delamination of the coating were observed. The non-continuity of the coating was probably caused by shrinkage of the EDB electrolyte material during sintering. Therefore, to heal the defects, a second cycle of deposition–sintering was carried out.

(**a**) (**b**)

Figure 7. Optical image of the surface of the sintered coating of the EDB electrolyte (sample EDB-1, Tsint = 850 °C, 5 h), obtained from a suspension in isopropanol/acetylacetone 70/30 vol.%, on a dense SDC substrate: (**a**) after EDB and sintering the first layer 6 μm thick; (**b**) after the second cycle of deposition–sintering of the EDB coating with a total thickness of 11 μm.

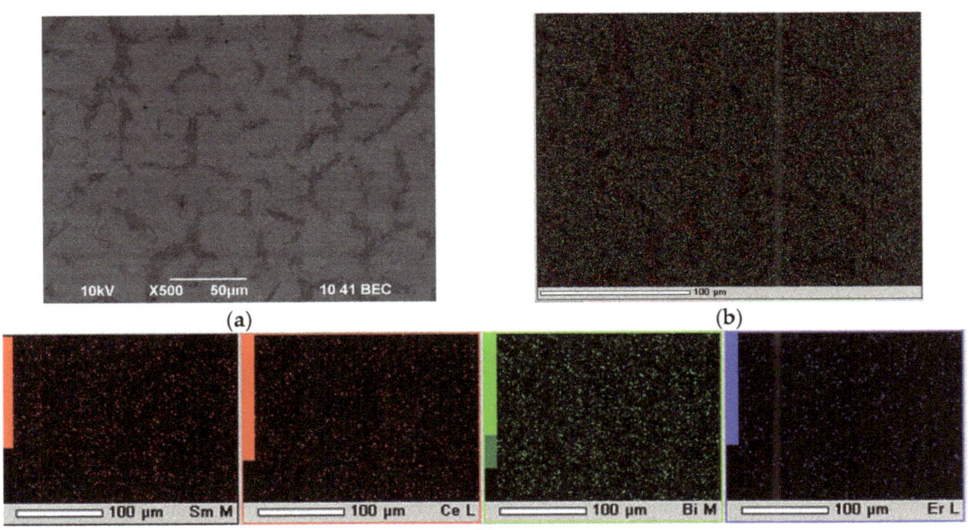

(**a**) (**b**) (**c**)

Figure 8. SEM image of the surface of the EDB coating (first deposition–sintering cycle) on the SDC substrate (EDB-2 sample, Tsint = 850 °C, 5 h) obtained using EPD from an EDB suspension in isopropanol with PEI and acetylacetone additives: (**a**) surface, ×500; (**b**) integral map of elements' distribution; (c) individual element's maps.

The second EDB layer was deposited in the same EPD mode, and a total coating thickness of 12 μm was obtained. The obtained EDB barrier layer on the EDB-2 sample did not contain breaks or cracks in the coating (Figure 9). This sample was used for the subsequent fabrication of a single button cell and its electrochemical characterization.

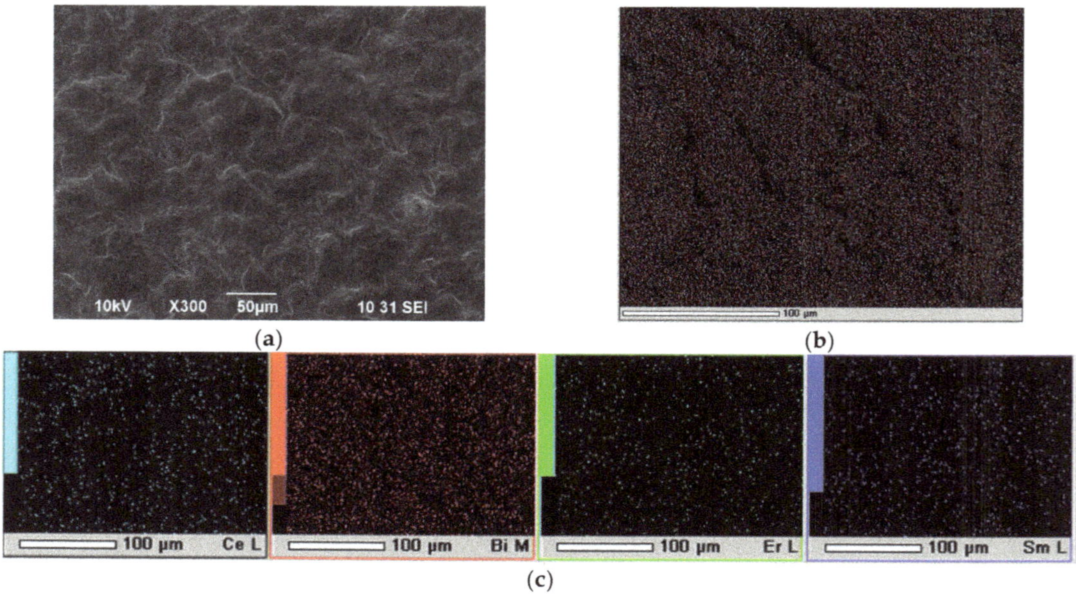

Figure 9. SEM image of the surface of the final EDB coating (two deposition–sintering cycles) on the SDC substrate (EDB-2 sample, Tsint = 850 °C, 5 h) obtained using EPD from EDB suspension in isopropanol medium with PEI and acetylacetone additives: (**a**) surface, ×300; (**b**) integral map of elements' distribution; (**c**) individual element maps.

3.5. Electrochemical Testing of Single SOFCs with a Supporting SDC Electrolyte Membrane and EDB Electrolyte Coating Applied Using EPD on the Cathode Side

For electrochemical characterization of the SDC electrolyte with an EDB barrier layer, a Pt/SDC/EDB/Pt cell was fabricated, as described in Section 2.4. The current–voltage characteristics were collected in the temperature range of 600–850 °C and the MPD values were calculated. Additionally, the polarization and ohmic (IR) resistance of the cell were measured using the current interruption method. The results are shown in Figure 10a,b, respectively. The main characteristic of the cell is the constant increase in the MPD values up to 850 °C, which is not typical for cells with supporting MIEC electrolytes. Usually, the cell performance decreases at temperatures higher than 800 °C due to the reduction of ceria in the anode conditions, which becomes significant at such high temperatures [5,48]. The MPD values are 12.6, 58.9, 101.2, 187.3, and 293.2 mW cm^{-2} at temperatures of 600, 700, 750, 800, and 850 °C, respectively. The obtained values are much higher than those obtained by Wachsman et al. in a cell with a supporting SDC electrolyte with an Er-doped Bi_2O_3 barrier layer and Pt/Au electrodes (51 mW cm^{-2} at 800 °C [14]) due to the reduction in the total electrolyte thickness from 800 to 562 μm. The OCV values obtained in this study were in the range of 725–750 mV and were close to those obtained for cells with doped Bi_2O_3 layers of similar thicknesses deposited with dip coating [16]. It should be noted that the polarization resistance of Pt electrodes was quite high and exceeded the ohmic resistance contribution even at such a high temperature as 750 °C. Due to the low sintering temperature established in this study for the EDB layer deposited using EPD, for future studies it would be preferable to use composite electrodes with Bi_2O_3 additives with reduced sintering temperatures that have been developed in various studies [49,50], including our own [51], which possess excellent electrochemical activity at decreased temperatures (polarization resistance as low as 0.6 Ω cm^2 at 600 °C [3]). This is important, as electrode polarization influences not only the cell performance, but also the OCV value. Thus, the replacement of the model Pt cathodes used in this study would help further improve the cell performance.

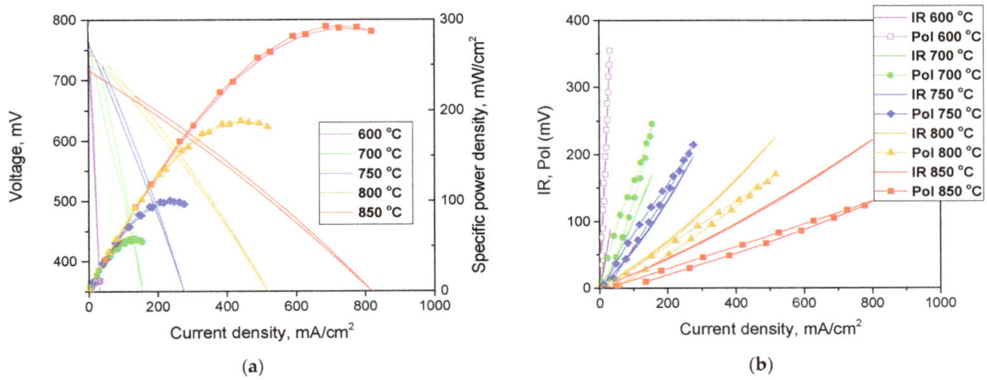

Figure 10. Electrochemical characterization of the Pt/SDC/EDB/Pt electrolyte-supported cell: (**a**) volt-ampere characteristics and cell power density; (**b**) ohmic and polarization cell resistances measured with the current interruption method.

To obtain information about the contributions of the electrolyte and the electrodes, the impedance of the cell was measured under air in both channels (designated as "Air") and under OCV conditions with air and wet hydrogen (3% H_2O) in the cathode and anode channels, respectively (designated as "SOFC mode"). The typical spectra are presented in Figure 11. Although the single-cell impedance was fairly complicated, the intercept of the impedance arc at the low-frequency side with the real axis usually embodies the overall cell resistance (R), while the inflection point on the impedance spectra at the high-frequency section represents the serial resistance (R_{hf}), which includes the electrolyte ohmic resistance as the main component, as well as the electrode/electrolyte interface resistance and the lateral resistance of the electrodes. Therefore, the difference between two intercepts represents the overall polarization resistance (R_p) from both the anode and cathode.

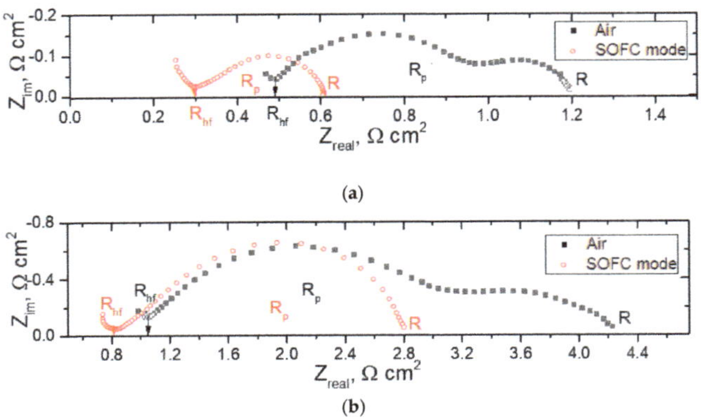

Figure 11. Typical spectra collected for the Pt/SDC/EDB/Pt electrolyte-supported cell collected in Aie and SOFC mode: (**a**) at 800 °C; (**b**) at 700 °C.

Table 3 summarizes the cell ohmic resistance and the polarization resistance of the electrodes, as well as the overall percentage of the ohmic resistance to the total resistance at different temperatures. The electrolyte ohmic resistance accounted for 44 and 49% of the total cell resistance at 850 °C, while at 600 °C it was only 3 and 16%. The presence of electronic conductivity in the electrolyte in the SOFC mode facilitates electrode reactions, thus decreasing R_p. Thus, as we mentioned above, at decreased temperatures, electrode polarization is the main deteriorating factor for the cell considered in this study. This agrees with data obtained using the current interruption method. Nevertheless, it should be noted that the R_p values of the cell presented in this study were significantly lower than those observed by Wachsman et al. [14], which ranged from 2 to 2.9 Ω cm^2 at 800 °C depending on the EDB thickness. This result is due to the electrode activation used in the study.

Table 3. Pt/SDC/EDB/Pt cell characteristics calculated from the EIS data.

Temperature	Air (Anode)/Air (Cathode)			Hydrogen (Anode)/Air (Cathode)		
	R_{hf}, Ω cm^2	R_p, Ω cm^2	$R_{hf}/(R_{hf} + R_p) \times 100$, %	R_{hf}, Ω cm^2	R_p, Ω cm^2	$R_{hf}/(R_{hf} + R_p) \times 100$, %
850	-	-	-	0.21	0.15	58
800	0.49	0.67	44	0.30	0.31	49
750	0.70	1.27	36	0.49	0.78	39
700	1.04	3.41	23	0.83	1.96	30
650	1.68	20.95	7	-	-	-
600	2.72	95.93	3	2.08	12.09	16

The electrolyte conductivity was calculated from the EIS data as follows:

$$\sigma_{el} = h/R_{hf}$$

where h is the total electrolyte thickness, including the supporting SDC electrolyte and the EDB coating. Figure 12 shows a comparison of the Arrhenius dependence of the conductivity with those obtained for SDC electrolytes without and with various barrier layers formed by EPD [9,15]. In the SOFC mode, the electrolyte conductivity is lower than that of SDC without electron-blocking layers and with a $BaCe_{0.8}Sm_{0.2}O_3$ + 1 wt. % CuO (BCS-CuO) anode barrier layer. However, it is similar to that of SDC with an Y-doped Bi_2O_3 coating, thus demonstrating the superior effectiveness of Bi-based cathodes to block electrons. In air, the SDC-EDB conductivity is higher than that of the SDC electrolyte due to the higher conductivity of EDB, and lower than that of BCS-CuO-SDC due to electronic conductivity in the BCS-CuO electrolyte under air conditions. The deposition of an EDB layer results in increasing the electrolyte conductivity; however, it does not change the activation energy of its conductivity (values are shown in Figure 12 using the relevant colors).

3.6. Microstructural Characterization of the SDC/EDB Electrolyte after Testing in the SOFC Mode

Figure 13 shows the SEM image of a Pt/SDC/EDB/Pt cell after testing in the SOFC mode. As shown in Figure 13a, the EDB layer has a dense sintered structure, and the layer thickness is in the range of 5–7 μm, which is lower than the calculated one. Good adhesion between the EDB layer and the SDC substrate should be noted, and this was maintained after three days of testing. Diffusion of bismuth cations was observed in the bulk of the SDC substrate. The inhomogeneity of the EDB layer may be related to its partial evaporation during the sintering process as well as baking of the Pt electrodes. However, the effect of the EDB layer manifested itself on the conductivity and power characteristics of the Pt/SDC/EDB/Pt cell (Figures 11 and 12). It should also be noted that the EDB layer obtained using EPD had a superior density compared with Bi-containing coatings obtained with screen-printing [15,20,52]. In general, obtaining dense Bi-containing layers is a challenging task, which can mainly be realized through complicated physical methods such as PLD and magnetron sputtering [20]. In this sense, a simple and inexpensive method of electrophoretic deposition offers undoubted advantages.

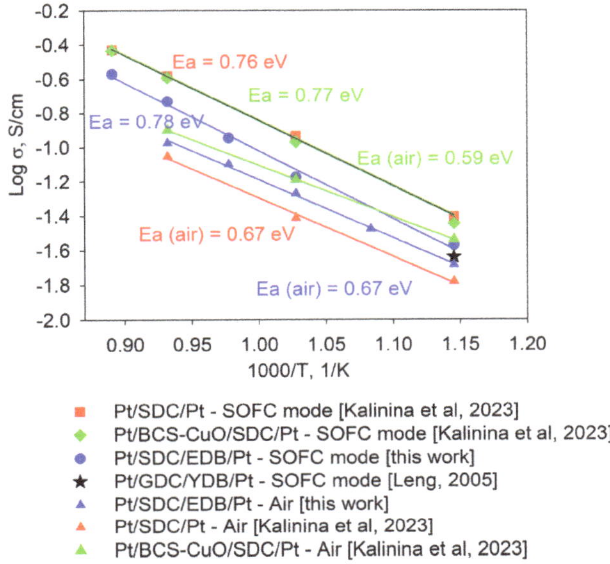

Figure 12. Arrhenius dependences (this work and [9,15]) of the electrolyte conductivity calculated from EIS data for the cells with the supporting SDC electrolyte with/without electron-blocking layers.

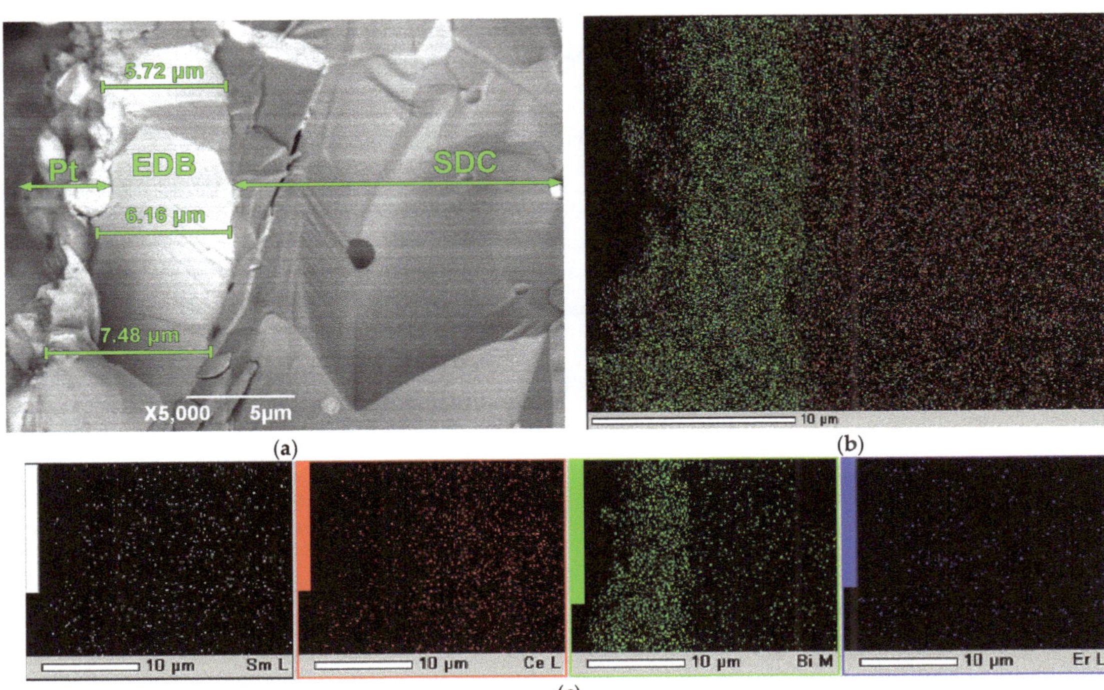

Figure 13. SEM image of a Pt/SDC/EDB/Pt cell after testing in the SOFC mode: (**a**) cleavage, ×5000. On the left, the Pt layer is shown, which was the electrode (cathode) during measurements in the SOFC mode; (**b**) integrated map of elements; (**c**) individual maps of elements.

4. Conclusions

In the present work, we studied the electrophoretic formation of an EDB electron-blocking coating on a supporting SDC electrolyte using a sublayer of a conductive polymer, polypyrrole. A variant of preparing a suspension of EDB powder based on isopropanol with small additions of polyethyleneimine (PEI, 0.26 g/L) and acetylacetone (0.15 g/L) is proposed to obtain continuous EPD coatings and eliminate cracking of the coatings after sintering at a temperature of 850 °C for 5 h. The electrokinetic properties of suspensions based on the EDB powder were studied. A non-sintered electrophoretically deposited EDB coating with a thickness of 12 µm was obtained; after sintering (850 °C, 5 h), the thickness of the EDB coating was 5–7 µm. The EDB coating was characterized by a dense sintered structure, the absence of pores, and good adhesion to the SDC substrate, including after testing in the SOFC mode for three days. Inhomogeneity in the thickness of the EDB coating after sintering was noted due to the partial evaporation of bismuth oxide. On the fabricated Pt/SDC/EDB/Pt SOFC cell, a monotonic increase in the maximum specific power with an increase in temperature up to 850 °C was noted. Therefore, the applied electron-blocking EDB layer decreased the effect of CeO_2 reduction from the anode side at high temperatures on the cell performance. The MPD values were 12.6, 58.9, 101.2, 187.3, and 293.2 mW cm^{-2} at temperatures of 600, 700, 750, 800, and 850 °C, respectively. The OCV values of the cell were not high; namely, they were in the range of 725–750 mV at temperatures of 600–850 °C. We suppose that such low values are due to high electrode polarization and can be improved by using electrodes that are highly active at lower temperatures. Measurements of the Pt/SDC/EDB/Pt cell showed that the conductivity in the SOFC mode decreased with respect to the Pt/SDC/Pt cell due to the blocking effect of the EDB layer. However, in air, the conductivity of the Pt/SDC/EDB/Pt cell was higher due to higher conductivity of the EDB layer. The results obtained within the framework of this study revealed that the EPD method can be considered as a promising technology for the formation of barrier layers based on highly conductive doped bismuth oxide, such as in SOFC technology, as well as in other application fields.

Author Contributions: Conceptualization, E.K. and E.P.; methodology, E.K.; software, E.P.; validation, E.K., L.E. and E.P.; investigation, E.K. and L.E.; resources, L.E.; data curation, E.K. and E.P.; writing—original draft preparation, E.K.; writing—review and editing, E.P.; visualization, E.K. and L.E.; supervision, E.P.; project administration, E.K. All authors have read and agreed to the published version of the manuscript.

Funding: This research received no external funding.

Institutional Review Board Statement: Not applicable.

Informed Consent Statement: Not applicable.

Data Availability Statement: The original contributions presented in this study are included in the article; further inquiries can be directed to the corresponding author.

Acknowledgments: The XRD studies were performed using the facilities of the Ural-M collective center (IMET UB RAS, Yekaterinburg, Russia). The authors are grateful to Sergey Pikalov (IMET UB RAS) and Elena Filonova (Ural Federal University, Yekaterinburg, Russia) for the XRD data treatment. The SEM studies were performed using the facilities of the ISSC UB RAS shared-access center (Yekaterinburg, Russia). The authors acknowledge Kirill Shubin (IHTE UB RAS, Yekaterinburg, Russia) for the electrochemical experiments.

Conflicts of Interest: The authors declare no conflict of interest.

References

1. Mendonça, C.; Ferreira, A.; Santos, D.M.F. Towards the Commercialization of Solid Oxide Fuel Cells: Recent Advances in Materials and Integration Strategies. *Fuels* **2021**, *2*, 393–419. [CrossRef]
2. Sreedhar, I.; Agarwal, B.; Goyal, P.; Singh, S.A. Recent Advances in Material and Performance Aspects of Solid Oxide Fuel Cells. *J. Electroanal. Chem.* **2019**, *848*, 113315. [CrossRef]

3. Singh, B.; Ghosh, S.; Aich, S.; Roy, B. Low Temperature Solid Oxide Electrolytes (LT-SOE): A Review. *J. Power Sources* **2017**, *339*, 103–135. [CrossRef]
4. Arunkumar, P.; Meena, M.; Babu, K.S. A Review on Cerium Oxide-Based Electrolytes for ITSOFC. *Nanomater. Energy* **2012**, *1*, 288–305. [CrossRef]
5. Pikalova, E.Y.; Kalinina, E.G. Solid Oxide Fuel Cells Based on Ceramic Membranes with Mixed Conductivity: Improving Efficiency. *Russ. Chem. Rev.* **2021**, *90*, 703–749. [CrossRef]
6. Azad, A.M.; Larose, S.; Akbar, S.A. Bismuth Oxide-Based Solid Electrolytes for Fuel Cells. *J. Mater. Sci.* **1994**, *29*, 4135–4151. [CrossRef]
7. Jeong, I.; Jeong, S.J.; Yun, B.-H.; Lee, J.-W.; Lee, C.-W.; Jung, W.; Lee, K.T. Physically Driven Enhancement of the Stability of Bi_2O_3-Based Ionic Conductors via Grain Boundary Engineering. *NPG Asia Mater.* **2022**, *14*, 53. [CrossRef]
8. Shri Prakash, B.; Pavitra, R.; Senthil Kumar, S.; Aruna, S.T. Electrolyte Bi-Layering Strategy to Improve the Performance of an Intermediate Temperature Solid Oxide Fuel Cell: A Review. *J. Power Sources* **2018**, *381*, 136–155. [CrossRef]
9. Kalinina, E.G.; Pikalova, E.Y. Electrophoretic Deposition of Dense Anode Barrier Layers of Doped ZrO_2 and $BaCeO_3$ on a Supporting $Ce_{0.8}Sm_{0.2}O_{2-\delta}$ Solid Electrolyte: Problems and Search for Solutions in SOFC Technology. *Int. J. Hydrog. Energy*, 2023; *in press*. [CrossRef]
10. Ling, Y.; Wang, X.; Ma, Z.; Wei, K.; Wu, Y.; Khan, M.; Zheng, K.; Shen, S.; Wang, S. Review of Experimental and Modelling Developments for Ceria-Based Solid Oxide Fuel Cells Free from Internal Short Circuits. *J. Mater. Sci.* **2020**, *55*, 1–23. [CrossRef]
11. Wachsman, E.D.; Lee, K.T. Lowering the Temperature of Solid Oxide Fuel Cells. *Science* **2011**, *334*, 935–939. [CrossRef]
12. Pesaran, A.; Jaiswal, A.; Ren, Y.; Wachsman, E.D. Development of a New Ceria/Yttria-Ceria Double-Doped Bismuth Oxide Bilayer Electrolyte Low-Temperature SOFC with Higher Stability. *Ionics* **2019**, *25*, 3153–3164. [CrossRef]
13. Concha-Balderrama, A.; Martinez-Rodriguez, H.A.; Rojas-George, G.; Esparza-Ponce, H.E.; Orozco-Carmona, V.; Pizá-Ruiz, P.; Bocanegra-Bernal, M.H.; Reyes-Rojas, A. Enhanced Ionic Transport and Compressive Residual Stress in Er-Doped Bi_2O_3 with Lower Er^{3+} Concentrations. *J. Electron. Mater.* **2018**, *47*, 5422–5432. [CrossRef]
14. Wachsman, E.D.; Jayaweera, P.; Jiang, N.; Lowe, D.M.; Pound, B.G. Stable High Conductivity Ceria/Bismuth Oxide Bilayered Electrolytes. *J. Electrochem. Soc.* **1997**, *144*, 233–236. [CrossRef]
15. Leng, Y.J. Development of YDB/GDC Composite Electrolyte For Low-Temperature Solid Oxide Fuel Cells. *ECS Proc. Vol.* **2005**, *2005–2007*, 1110–1116. [CrossRef]
16. Park, J.-Y.; Wachsman, E.D. Stable and High Conductivity Ceria/Bismuth Oxide Bilayer Electrolytes for Lower Temperature Solid Oxide Fuel Cells. *Ionics* **2006**, *12*, 15–20. [CrossRef]
17. Duan, N.; Ma, J.; Li, J.; Yan, D.; Chi, B.; Pu, J.; Li, J. High Performance Cathode-Unsintered Solid Oxide Fuel Cell Enhanced by Porous $Bi_{1.6}Er_{0.4}O_3$ (ESB) Interlayer. *Int. J. Hydrog. Energy* **2018**, *43*, 12713–12719. [CrossRef]
18. Lee, J.G.; Park, M.G.; Yoon, H.H.; Shul, Y.G. Application of GDC-YDB Bilayer and LSM-YDB Cathode for Intermediate Temperature Solid Oxide Fuel Cells. *J. Electroceram.* **2013**, *31*, 231–237. [CrossRef]
19. Zhang, L.; Xia, C.; Zhao, F.; Chen, F. Thin Film Ceria–Bismuth Bilayer Electrolytes for Intermediate Temperature Solid Oxide Fuel Cells with $La_{0.85}Sr_{0.15}MnO_{3-\delta}$–$Y_{0.25}Bi_{0.75}O_{1.5}$ Cathodes. *Mater. Res. Bull.* **2010**, *45*, 603–608. [CrossRef]
20. Ahn, J.S.; Camaratta, M.A.; Pergolesi, D.; Lee, K.T.; Yoon, H.; Lee, B.W.; Jung, D.W.; Traversa, E.; Wachsman, E.D. Development of High Performance Ceria/Bismuth Oxide Bilayered Electrolyte SOFCs for Lower Temperature Operation. *J. Electrochem. Soc.* **2010**, *157*, B376. [CrossRef]
21. Pikalova, E.Y.; Kalinina, E.G. Place of Electrophoretic Deposition among Thin-Film Methods Adapted to the Solid Oxide Fuel Cell Technology: A Short Review. *Int. J. EQ* **2019**, *4*, 1–27. [CrossRef]
22. Pikalova, E.Y.; Kalinina, E.G. Electrophoretic Deposition in the Solid Oxide Fuel Cell Technology: Fundamentals and Recent Advances. *Renew. Sust. Energy Rev.* **2019**, *116*, 109440. [CrossRef]
23. Kalinina, E.G.; Pikalova, E.Y. New Trends in the Development of Electrophoretic Deposition Method in the Solid Oxide Fuel Cell Technology: Theoretical Approaches, Experimental Solutions and Development Prospects. *Russ. Chem. Rev.* **2019**, *88*, 1179–1219. [CrossRef]
24. Guo, X.; Li, X.; Lai, C.; Li, W.; Zhang, D.; Xiong, Z. Cathodic Electrophoretic Deposition of Bismuth Oxide (Bi_2O_3) Coatings and Their Photocatalytic Activities. *Appl. Surf. Sci.* **2015**, *331*, 455–462. [CrossRef]
25. Liang, T.; Guo, X. Remarkably Facile Preparation of Superhydrophobic Functionalized Bismuth Trioxide (Bi_2O_3) Coatings. *Appl. Sci.* **2019**, *9*, 2653. [CrossRef]
26. Megnin, C. Electrophoretic Deposition of $BiVO_4$ Layers on FTO Substrates for Photo Electro-chemical Cells. *Ceramics-Silikaty* **2019**, *63*, 124–130. [CrossRef]
27. Chahkandi, M.; Zargazi, M. New Water Based EPD Thin $BiVO_4$ Film: Effective Photocatalytic Degradation of Amoxicillin Antibiotic. *J. Hazard. Mater.* **2020**, *389*, 121850. [CrossRef]
28. Joh, D.W.; Park, J.H.; Kim, D.Y.; Yun, B.-H.; Lee, K.T. High Performance Zirconia-Bismuth Oxide Nanocomposite Electrolytes for Lower Temperature Solid Oxide Fuel Cells. *J. Power Sources* **2016**, *320*, 267–273. [CrossRef]
29. Park, J.W.; Joh, D.W.; Yun, B.-H.; Samdani, K.J.; Lee, K.T. Development of Nanostructured $La_{0.8}Sr_{0.2}MnO_{3-\delta}$-$Er_{0.4}Bi_{1.6}O_3$ Cathodes via an Infiltration Process with Different Polymeric Agents for Intermediate Temperature Solid Oxide Fuel Cells. *Int. J. Hydrog. Energy* **2017**, *42*, 6332–6337. [CrossRef]
30. Wang, Z.; Sun, H.; Li, J.; Guo, X.; Hu, Q.; Yang, Z.; Yu, F.; Li, G. Modified $La_{0.6}Sr_{0.4}Co_{0.2}Fe_{0.8}O_{3-\delta}$ Cathodes with the Infiltration of $Er_{0.4}Bi_{1.6}O_3$ for Intermediate-Temperature Solid Oxide Fuel Cells. *Int. J. Hydrog. Energy* **2021**, *46*, 22932–22941. [CrossRef]

31. Sadykov, V.; Usoltsev, V.; Yeremeev, N.; Mezentseva, N.; Pelipenko, V.; Krieger, T.; Belyaev, V.; Sadovskaya, E.; Muzykantov, V.; Fedorova, Y.; et al. Functional Nanoceramics for Intermediate Temperature Solid Oxide Fuel Cells and Oxygen Separation Membranes. *J. Europ. Ceram. Soc.* **2013**, *33*, 2241–2250. [CrossRef]
32. Li, J.; Wang, S.; Wang, Z.; Liu, R.; Wen, T.; Wen, Z. $La_{0.84}Sr_{0.16}MnO_{3-\delta}$ Cathodes Impregnated with $Bi_{1.4}Er_{0.6}O_3$ for Intermediate-Temperature Solid Oxide Fuel Cells. *J. Power Sources* **2009**, *194*, 625–630. [CrossRef]
33. Mukasyan, A.S.; Epstein, P.; Dinka, P. Solution Combustion Synthesis of Nanomaterials. *Proc. Combust. Inst.* **2007**, *31*, 1789–1795. [CrossRef]
34. Rodríguez-Carvajal, J. Recent Developments of the Program FULLPROF, in Commission on Powder Diffraction (IUCr). *Newsletter* **2001**, *26*, 12–19. Available online: http://journals.iucr.org/iucr-top/comm/cpd/Newsletters (accessed on 1 January 2023).
35. Kalinina, E.; Shubin, K.; Pikalova, E. Electrophoretic Deposition and Characterization of the Doped $BaCeO_3$ Barrier Layers on a Supporting $Ce_{0.8}Sm_{0.2}O_{1.9}$ Solid-State Electrolyte. *Membranes* **2022**, *12*, 308. [CrossRef]
36. Chiodelli, G.; Malavasi, L. Electrochemical Open Circuit Voltage (OCV) Characterization of SOFC Materials. *Ionics* **2013**, *19*, 1135–1144. [CrossRef]
37. Cobaner, S.; Yilmaz, S. Electrical and Structural Properties of New Type Er and Yb Doped Bismuth Oxide Solid Electrolytes Synthesized by Pechini Method. *J. Electroceram.* **2021**, *46*, 83–92. [CrossRef]
38. Arasteh, S.; Maghsoudipour, A.; Alizadeh, M.; Nemati, A. Effect of Y_2O_3 and Er_2O_3 Co-Dopants on Phase Stabilization of Bismuth Oxide. *Ceram. Int.* **2011**, *37*, 3451–3455. [CrossRef]
39. Kirillov, S.A. Surface Area and Pore Volume of a System of Particles as a Function of Their Size and Packing. *Micropor. Mesopor. Mater.* **2009**, *122*, 234–239. [CrossRef]
40. Kalinina, E.; Ivanov, M. The Electrophoretic Deposition of Nanopowders Based on Yttrium Oxide for Bulk Ceramics Fabrication. *Inorganics* **2022**, *10*, 243. [CrossRef]
41. Hosomi, T.; Matsuda, M.; Miyake, M. Electrophoretic Deposition for Fabrication of YSZ Electrolyte Film on Non-Conducting Porous NiO–YSZ Composite Substrate for Intermediate Temperature SOFC. *J. Europ. Ceram. Soc.* **2007**, *27*, 173–178. [CrossRef]
42. Azarian Borojeni, I.; Raissi, B.; Maghsoudipour, A.; Kazemzad, M.; Talebi, T. Fabrication of Solid Oxide Fuel Cells (SOFCs) Electrolytes by Electrophoretic Deposition (EPD) and Optimizing the Process. *KEM* **2015**, *654*, 83–87. [CrossRef]
43. Suarez, G.; Nguyen, N.T.K.; Rendtorff, N.M.; Sakka, Y.; Uchikoshi, T. Electrophoretic Deposition for Obtaining Dense Lanthanum Silicate Oxyapatite (LSO). *Ceram. Inter.* **2016**, *42*, 19283–19288. [CrossRef]
44. Kalinina, E.G.; Pikalova, E.Y. Formation of a Single- and Two-Layer Solid Electrolyte by Electrophoresis on Anodic Substrates Metalized with Silver or Platinum. *Russ. J. Phys. Chem.* **2022**, *96*, 2763–2773. [CrossRef]
45. Suzuki, H.T.; Uchikoshi, T.; Kobayashi, K.; Suzuki, T.S.; Sugiyama, T.; Furuya, K.; Matsuda, M.; Sakka, Y.; Munakata, F. Fabrication of GDC/LSGM/GDC Tri-Layers on Polypyrrole-Coated NiO-YSZ by Electrophoretic Deposition for Anode-Supported SOFC. *J. Ceram. Soc. Jpn.* **2009**, *117*, 1246–1248. [CrossRef]
46. Das, D.; Basu, R.N. Electrophoretic Deposition of Zirconia Thin Film on Nonconducting Substrate for Solid Oxide Fuel Cell Application. *J. Am. Ceram. Soc.* **2014**, *97*, 3452–3457. [CrossRef]
47. Uchikoshi, T.; Furumi, S.; Shirahata, N.; Suzuki, T.S.; Sakka, Y. Conductive Polymer Coating on Nonconductive Ceramic Substrates for Use in the Electrophoretic Deposition Process. *J. Am. Ceram. Soc.* **2008**, *91*, 1674–1677. [CrossRef]
48. Jaiswal, N.; Tanwar, K.; Suman, R.; Kumar, D.; Upadhyay, S.; Parkash, O. A Brief Review on Ceria Based Solid Electrolytes for Solid Oxide Fuel Cells. *J. Alloys Compd.* **2019**, *781*, 984–1005. [CrossRef]
49. Jung, W.C.; Chang, Y.-J.; Fung, K.-Z.; Haile, S. High Electrochemical Activity of Bi_2O_3-Based Composite SOFC Cathodes. *J. Korean Ceram. Soc.* **2014**, *51*, 278–282. [CrossRef]
50. Fang, W.; Yang, T.; Huang, K. In Situ Synthesis of a High-Performance Bismuth Oxide Based Composite Cathode for Low Temperature Solid Oxide Fuel Cells. *Chem. Commun.* **2019**, *55*, 2801–2804. [CrossRef]
51. Pikalova, E.; Bogdanovich, N.; Kolchugin, A.; Shubin, K.; Ermakova, L.; Eremeev, N.; Farlenkov, A.; Khrustov, A.; Filonova, E.; Sadykov, V. Development of Composite $LaNi_{0.6}Fe_{0.4}O_{3-\delta}$-Based Air Electrodes for Solid Oxide Fuel Cells with a Thin-Film Bilayer Electrolyte. *Int. J. Hydrog. Energy* **2021**, *46*, 16947–16964. [CrossRef]
52. Leng, Y.J.; Chan, S.H. Anode-Supported SOFCs with Y_2O_3-Doped Bi_2O_3/Gd_2O_3-Doped CeO_2 Composite Electrolyte Film. *Electrochem. Solid-State Lett.* **2006**, *9*, A56–A59. [CrossRef]

coatings

MDPI

Communication

Electrodeposition of Soft Magnetic Fe-W-P Alloy Coatings from an Acidic Electrolyte

Natalia Kovalska [1,2,3,*], Antonio Mulone [4], Jordi Sort [5,6], Uta Klement [4], Gurdial Blugan [3], Wolfgang Hansal [2] and Wolfgang Kautek [1,*]

1 Department of Physical Chemistry, University of Vienna, A-1090 Vienna, Austria
2 RENA Technologies Austria GmbH, A-2700 Wiener Neustadt, Austria; wolfgang.hansal@elektrochemie.eu
3 Laboratory for High Performance Ceramics, Empa, Swiss Federal Laboratories for Materials Science & Technology, CH-8600 Dübendorf, Switzerland
4 Department of Industrial and Materials Science, Chalmers University of Technology, SE-41296 Gothenburg, Sweden; mulone@chalmers.se (A.M.); uta.klement@chalmers.se (U.K.)
5 Department de Física, Facultat de Ciències, Universitat Autònoma de Barcelona, E-08193 Bellaterra, Spain; sort.jordi@gmail.com
6 Institució Catalana de Recerca i Estudis Avançats (ICREA), E-08010 Barcelona, Spain
* Correspondence: natalia.kovalska@empa.ch (N.K.); wolfgang.kautek@univie.ac.at (W.K.)

Abstract: Fe-W-P coatings were deposited from a newly developed electrolytic bath. The effect of plating parameters, such as electrolyte current density and pH has been studied. It was found that the pH has a very strong effect on the phosphorous content of the coatings. Metallic-like, non-powdery alloys of Fe-W-P deposits with no cracks (lowly stressed) can be obtained at a lower pH (<3), exhibiting high phosphorus (up to 13 at.%) and low tungsten (6 at.%) contents. At a higher pH (>3), the composition changes to low phosphorus and high tungsten content, showing a matte, greyish, and rough surface. The applied current density also influences the morphology and the amount of phosphorous content. The deposits showed an amorphous structure for all samples with soft ferromagnetic properties.

Keywords: electrodeposition; Fe-W-P alloy; hydrogen evolution; pH; soft magnetic

Citation: Kovalska, N.; Mulone, A.; Sort, J.; Klement, U.; Blugan, G.; Hansal, W.; Kautek, W. Electrodeposition of Soft Magnetic Fe-W-P Alloy Coatings from an Acidic Electrolyte. *Coatings* **2023**, *13*, 801. https://doi.org/10.3390/coatings13040801

Academic Editors: Gergana Alexieva and Qi Hua Fan

Received: 13 March 2023
Revised: 14 April 2023
Accepted: 16 April 2023
Published: 20 April 2023

1. Introduction

The current trends towards the reduction of hazardous procedures and cost competitiveness become dominant in many new applications such as in micro-electromechanical systems (MEMS) and nano-electromechanical systems (NEMS). Amorphous iron group coatings with phosphorous or/and tungsten are of industrial interest due to their soft magnetic behaviour, application in recording media, sensors, data storage media, inductive devices, and transformer cores [1–6]. It has been shown that the addition of phosphorous can improve the magnetic properties of electrodeposited coatings [7,8]. A detailed study based on electrochemical in-situ techniques led to a mechanistic understanding of the Fe-P alloy electrodeposition [9]. An increase of the P content exerts a beneficial effect on the corrosion resistance [10,11].

The addition of P in Co-W and Fe-W coatings results in compact and dense layers with decreased grain size [12–16]. The presence of W in Fe-P alloys significantly improves properties such as hardness, heat stability, wear resistance, corrosion protection, and magnetic characteristics [15,17–20]. This Fe-W-P coating can compete with the high-cost Ni-W-P alloy used in decorative finishing, and can be considered as an alternative to Cr, which is produced in the hazardous process based on Cr(VI) [21].

The electrodeposition of W can only take place in the presence of iron group metals through an induced co-deposition [13,22]. Most Fe-W-P alloys studied so far exhibit the presence of a higher amount of W and a low P content. These coatings are generated from electrolytes containing $Na_2WO_4 \cdot 2H_2O$, as a W source at a moderate alkaline pH (~8). Some authors obtained Fe-W alloys by means of electrodeposition in citrate-containing solutions [7,8,17,18,22]. A generally accepted mechanism for induced co-deposition of W with iron group metals (M) in the presence of citrate implies the formation of M(II)-citrate complexes, which favours tungstate reduction to adsorbed species composed of W oxides with citrate complexes [21,22].

Considering the numerous advantages of Fe-P coatings with high P content [9–11], the aim of this work is to explore a new electrolytic bath using sustainable or low quantity hazardous materials, which leads to an increase in the P content in Fe-W-P alloys. Previous studies reported the formation of alloys, where the P content dominates over W. Therefore, we describe the use of a new source of elements and the influence of the chemical environment on the final alloy composition.

2. Materials and Methods

The electrolyte components for Fe-W-P electrodeposition are shown in Table 1. Analytical grade chemicals and deionized water were used. The pH was adjusted to 3 and 4 by the addition of H_2SO_4 and NaOH. All electrodeposition experiments were carried out at 60 °C.

Table 1. Bath composition.

Components	Molarity/M	Concentration/gL^{-1}
$FeSO_4 \cdot 7H_2O$	0.72	200
$H_2PO_4 \cdot 12WO_3 \cdot H_2O$	0.02	60
Citrate	0.39	100
Glycine	0.64	48
$NaH_2PO_2 \cdot H_2O$	0.08	7
$(NH_4)_2SO_4$	0.45	60

The Electrochemical Quartz Microbalance (EQMB) measurements were performed with a three-electrode system equipped with an EQCM module (Autolab PGSTAT, Metrohm, Filderstadt, Germany). 6 MHz AT-cut quartz crystals were coated with 100 nm thick gold layers. An area of 0.384 cm^2 was exposed. A gold coil was used as a counter electrode and a Ag/AgCl/3M KCl electrode was used as a reference electrode (+0.21 V vs. standard hydrogen electrode, SHE) [23]. For easier data evaluation the potentials are presented versus the SHE as U_{SHE}. The gold surface was cleaned by cycling in 0.1 M H_2SO_4 solution.

Galvanostatic deposition experiments were performed in a beaker of 0.3 l without stirring. A brass plate served as a substrate. All samples were rinsed in 0.1 M H_2SO_4 before deposition. The thickness of the electrodeposited coatings varied between 11 and 15 μm. The thickness was calculated based on weight elemental analyses of the electrodeposited alloys [24]. The reference electrode was Ag/AgCl/3M KCl (+0.21 V vs. standard hydrogen electrode, SHE) [23]. As the counter electrode served a steel plate (SAE 304). It exposed ~30 cm^2 to the electrolyte and was positioned parallel to the working electrode at a distance of ~10 cm. The current efficiency of the alloy deposition was determined by Faraday's law [25].

The coating surfaces were characterized with a scanning electron microscope (Hitachi FEG-SEM S4800, Tokyo, Japan) equipped with EDX for elemental analysis. An acceleration voltage of 5 kV was used for the imaging mode, and a voltage of 20 kV for the EDX mode. The vacuum pressure in the specimen chamber was of the order of 10^{-4} Pa. Crystallographic information was obtained by means of an X-ray diffractometer (Bruker AXS D8, Karlsruhe, Germany) operated with Cr Kα radiation (λ = 2.0821 Å) at 35 kV and 40 mA, respectively.

Hysteresis loops were performed at room temperature using a vibrating sample magnetometer (EZ9 Microsense, LOT Quantum Design, Darmstadt, Germany) with a maximum magnetic field of 20 kOe. The loops were measured applying the field along the in-plane direction of the films.

3. Results and Discussions

The development of the presently investigated Fe-W-P electrolytic bath was based on electrolytes for the deposition of semi-amorphous, smooth, porous free Fe-P alloys [11]. To obtain a stable Fe-W-P electrolyte for long-term use, an acidic solution with the tungsten source phosphotungstic acid (PTA; $H_2PO_4 \cdot 12WO_3 \cdot H_2O$) was chosen in the present study, which had not been used before. The conjugate base of this heteropoly acid is the oxyanion $PW_{12}O_{40}{}^{3-}$, where W has an oxidation state of +6 [26]. PTA decomposes with increasing the pH [26,27].

3.1. Electrochemical Quartz Microbalance

The Fe-W-P alloy co-deposition was studied by the electrochemical quartz microbalance (EQMB). The measurements were performed to evaluate the influence of pH on the Fe-W-P deposition, compared to the Fe-W alloy deposition (Figure 1). The standard potentials of the Fe/Fe^{2+} electrode ($U° = -0.44$ V [28]) and the phosphorous deposition either from hypophosphate (oxidation state +1) [29],

$$H_3PO_2 + H^+ + e^- \rightleftharpoons P + H_2O \ (U° = -0.51 \text{ V}), \quad (1)$$

or from phosphotungstic acid (oxidation state +5) [29],

$$H_3PO_4 + 5H^+ + 5e^- \rightleftharpoons P + 4H_2O \ (U° = -0.40 \text{ V}) \quad (2)$$

are close to each other.

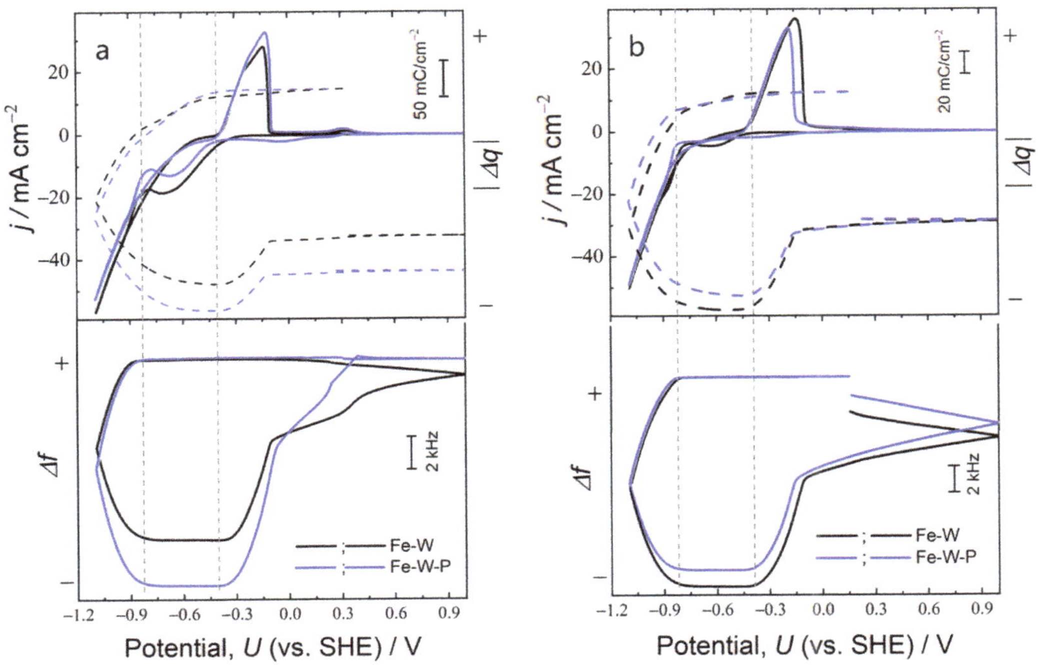

Figure 1. Electrochemical quartz microbalance data of the Fe-W and Fe-W-P electrodeposition (third potential cycle): current and charge density, and frequency change. (**a**) pH 3 and (**b**) pH 4.

An EDX analysis of a deposit from an electrolyte without hypophosphate ($NaH_2PO_2 \cdot H_2O$) resulted in practically zero (0.2–0.3 at.%) P content, even though phosphate was present in the form of $H_2PO_4 \cdot 12WO_3 \cdot H_2O$ (comp. Table 1). In this case, the W content indicated 6–8 at %.

A negative potential scan in the Fe-W and the Fe-W-P baths at pH 3 resulted in a negative current wave at ca. −0.4 V (Figure 1a), which can be interpreted as hydrogen evolution supported by the measured charge density $|\Delta q|$ [9]. The absence of any frequency change (Δf), i.e., any mass increase, indicates the generation of a gas uncoupled from the surface. With a negative potential of −0.85 V, co-deposition of the alloy components is indicated by a mass and current increase. The phosphotungstic acid contains W in the oxidation state of +6 (comp. WO_3). Its reduction to elementary tungsten occurs via the following partial steps [30]:

$$WO_3 + 2H^+ + 2e^- \rightleftharpoons WO_2 + H_2O\ (U^\circ = -0.036\ V) \tag{3}$$

$$WO_2 + 4H^+ + 4e^- \rightleftharpoons W + 2H_2O\ (U^\circ = -0.119\ V) \tag{4}$$

The respective Nernst potential of the W formation Equation (4) at pH 3 and 4, respectively, is approximately −0.8 V, which is close to the appearance potential of the Fe-W-P alloy (Figure 1). The hydrogen evolution is enhanced by lowering the pH, as seen by the increase of the *j*-wave and the charge increase negative of ca. −0.40 V.

The anodic stripping peaks of the alloys exhibit an asymmetry indicating the formation of a passive layer, most probably due to tungsten oxides, so that the potentiostat cannot compensate for the IR drop between the working and the reference electrodes. This phenomenon has not been observed without W [9]. After the abrupt decay of the anodic stripping peak, approximately 30–40% of the alloy mass remained. In the course of the further anodic potential scan this mass is further reduced practically without any charge transfer due to the chemical dissolution (hydrolysis) of the passive W phases [31], e.g.,

$$WO_{3(s)} + H_2O \rightarrow H_2WO_{4(aq)} \tag{5}$$

or

$$WO_{3(s)} + OH^- \rightarrow WO_2{}^{4-}{}_{(aq)} + H^+ \tag{6}$$

3.2. *Influence of pH, P and W on the Coatings*

A composition analysis was performed on the electrodeposited Fe-W-P coatings. The P and W content varies by changing the pH, as shown in Figure 2a. The current density shows a minor influence except in the case of P at low *j*. The reduction of pH from 4 to 3 leads to an increase in P and W contents [32]. At a higher H^+ concentration, the solubility of $H_3PO_{2(aq)}$ and $H_3PO_{4(aq)}$ species is higher [33] supporting a higher P deposition rate. Moreover, an increase in H^+ supports the reduction reaction of both P and W according to Equations (1), (3) and (4). The current efficiency (η) is higher in a less acidic electrolyte, i.e., at pH 4, and is more or less unaffected by the change in current density (Figure 2b). This can be explained by a lower hydrogen evolution competing with the alloy deposition (comp. Figure 1b). In case of an electrolytic bath at pH 3, the current efficiency increases with higher current density [32]. The lowest current efficiency is observed at a current density of 30 mA cm^{-2} and pH 3 (Figure 2b) resulting in a slow deposition rate of the alloy and a maximum P content of ca. 13 at.% (Figure 2a).

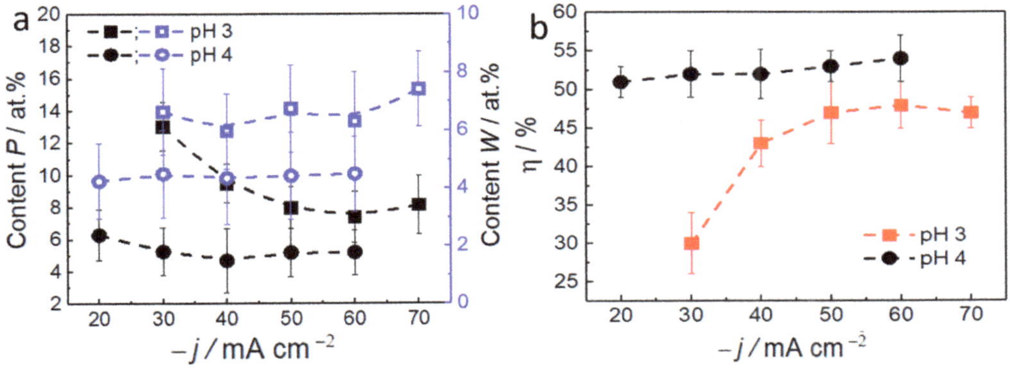

Figure 2. (**a**) Dependence of the P (black and solid) and W (blue and open) content in Fe-W-P coatings on the electrodeposition current density *j* at two acidic pH values. (**b**) Dependence of the current efficiency η on the electrodeposition current density *j* of the Fe-W-P electrodeposition at two acidic pH values.

The SEM images of the alloy show the effect of the pH and current density (Figure 3). The observed layer morphology is controlled by the competition between hydrogen bubble evolution, bubble detachment, and the layer growth. pH and *j* control η [32] (Figure 2b). A low η value indicates a high rate of H^+ reduction, a vigorous hydrogen gas evolution, and therefore a fast bubble detachment [34].

At pH 4, η is relatively high and practically independent of *j*. The H_2-bubble formation and detachment are relatively slow, so that an overgrowth of the rapidly depositing Fe-W-P alloy leads to partially cracked film bumps which appear hollow (Figure 3). A powdery dark greyish coating character is observed by the naked eye. Additionally, the SEM resolution is limited due to charging of the low conductivity deposit. This phase with low P content (Figure 2) exhibits an enhanced corrosion/oxidation rate [10,11] so that its conductivity drops.

At pH3, the bumps are not cracked. η is relatively low and the bubble formation and detachment is extremely vigorous. Therefore, the overgrowth of attached H_2-bubbles could not take place. The good resolution of the SEM images at all magnifications indicates low charging, suggesting a high conductivity of the coating with an obviously low oxidic conversion overlayer. The naked eye aspect is glossy and not powdery. For this reason, this coating type (pH3) is further investigated in the following.

The elements' partial current distribution, j_e, of the corresponding element "e", i.e., Fe, W, and P, were determined at pH 3 (Figure 4). The partial current densities were estimated from the EDX data and the weighed coating mass [11]. j_e as a function of the overall current density, shows that the current efficiency η of electrodeposition was increasing with *j*. j_e of the hydrogen evolution (side reaction) is much less affected by *j* than all other alloy components [34]. The exceptionally high P content at 30 mA cm^{-2} and pH3 (Figure 2a) can be understood on the basis of the evaluation in Figure 4. The relative deposition rate of Fe is much lower than that of P and W. At $j > 30$ mA cm^{-2}, all elements show a practically similar deposition rate increase with *j*.

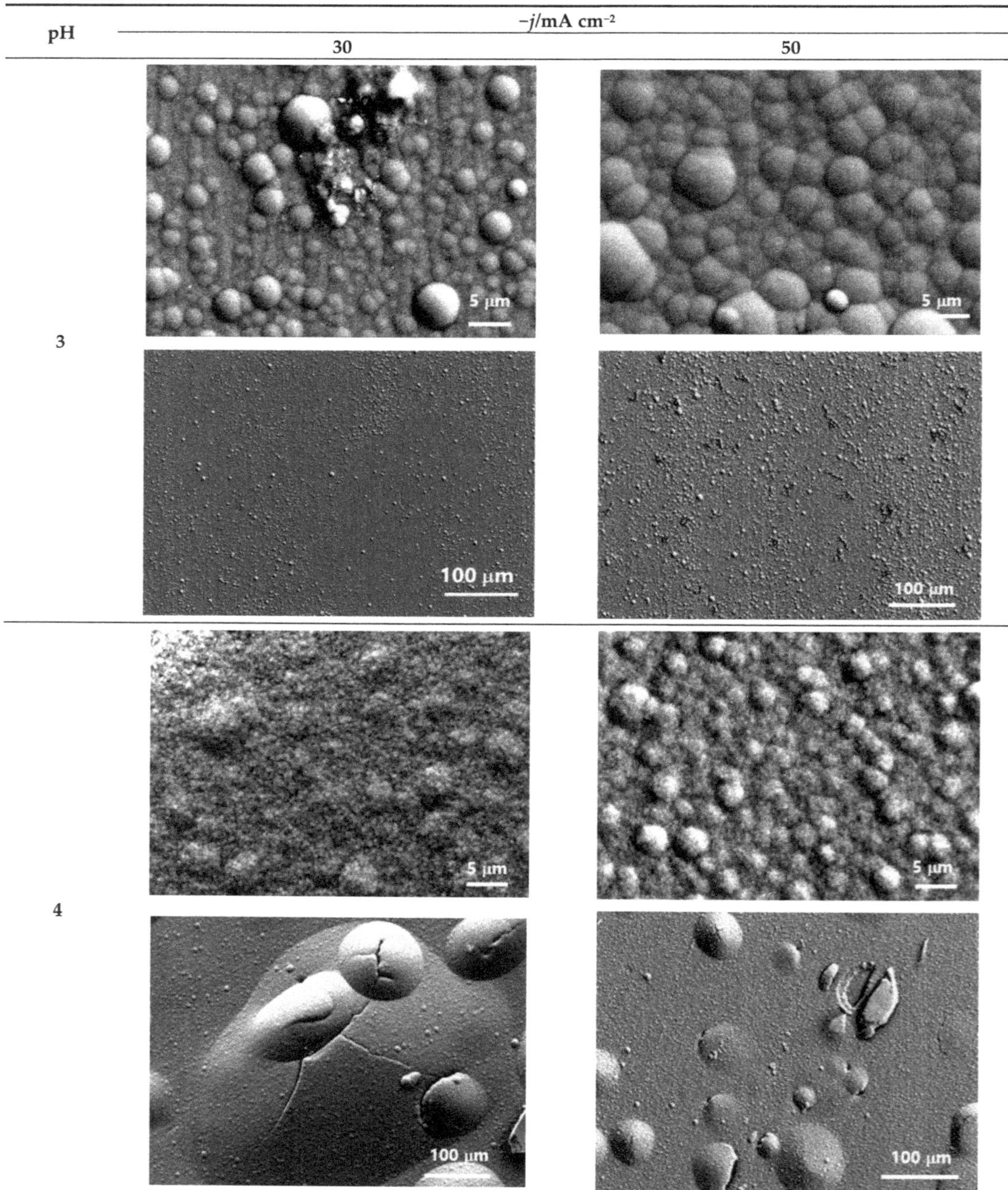

Figure 3. SEM morphology of Fe-W-P coatings deposited with DC electrodeposition at pH 3 and 4, respectively, and current density 30 and 50 mA cm^{-2}. Images presented with different magnifications.

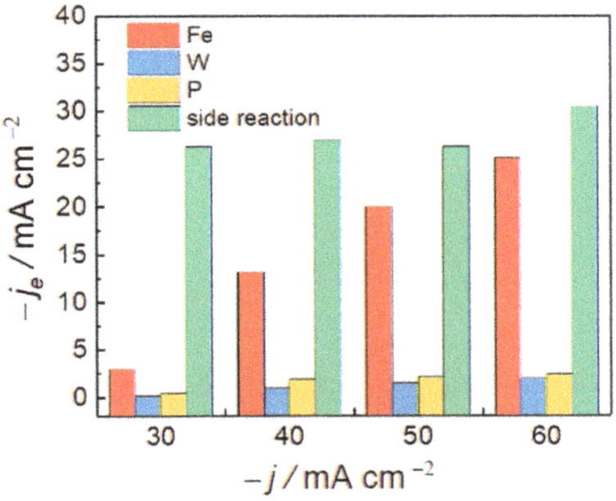

Figure 4. Partial deposition current densities j_e of the elements Fe, W, P, and the hydrogen evolution (side reaction), j vs. the applied current densities. pH 3.

XRD results of the Fe-W-P coatings are presented in Figure 5. The three samples show an amorphous structure, as indicated by the presence of a broad peak from 65° to 75°. This peak appears towards lower angles with respect to the 2θ (110) reflection characteristic of bcc-iron (i.e., 68.78°). This is due to the W substitution in the bcc-iron lattice, resulting in an increased lattice parameter [20]. The only visible crystalline peak corresponds to the brass substrate. The varying intensities of the brass peak for the various at.% P are due to the respective coating thicknesses.

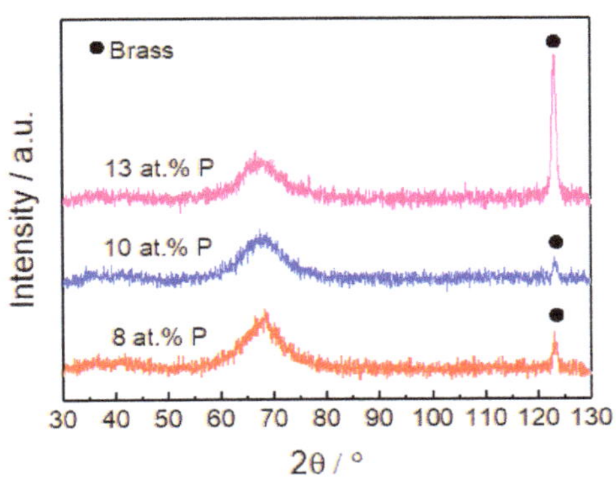

Figure 5. X-ray diffractograms of the Fe-W-P coatings deposited with increasing P content.

3.3. Magnetic Properties

Previous studies showed that the amounts of W and P in this type of alloy have an influence on the magnetic properties of the coatings [18,35–37]. In particular, the coatings are soft magnetic (with almost negligible coercivity) when the W or P contents are beyond 20 at.%. In the present study, magnetic measurements were performed applying the magnetic field in the plane of the films (Figure 6). The obtained results reveal that the Fe-W-P alloys are soft magnetic with coercively values lower than 25 Oe, although the percentages of the non-metallic element P in the alloy compositions were lower than 20 at.%. This is probably due to a negligible magnetocrystalline anisotropy, as was reported for other amorphous or nanocrystalline alloys containing a non-ferromagnetic element [38,39]. The addition of non-metallic elements P to Fe and W also leads to a reduction of the saturation magnetization, M_s [35].

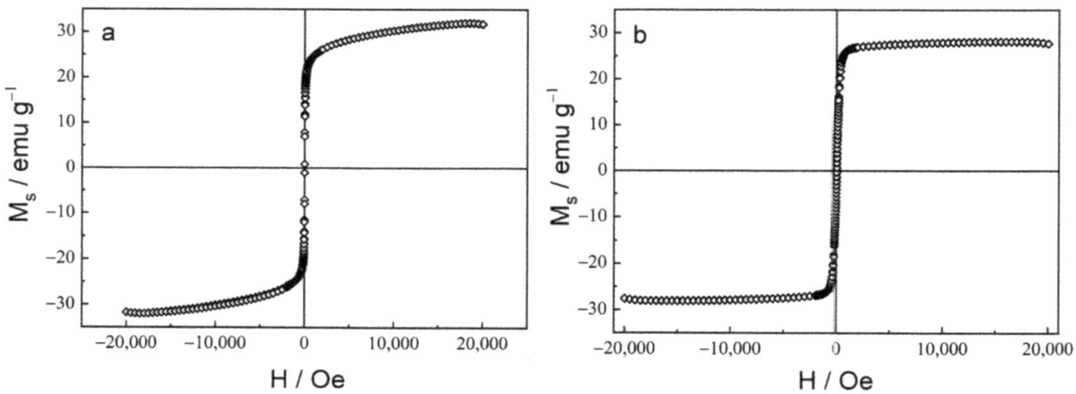

Figure 6. Hysteresis loops of Fe-W-P coatings. (**a**) 8 at.% P, 6 at.% W; (**b**) 13 at.% P, 6 at.% W, recorded with an external magnetic field applied along the in-plane direction.

For example, Figure 6 compares two alloys with the same amount of W (6 at.%) and different P contents. All deposited samples exhibit low ferromagnetic signals, i.e., M_s = 20−40 emug^{-1}. The reason for such low M_s could be that a mixture of non-ferromagnetic and weakly ferromagnetic regions co-exist in the coatings [35]. The increase of the P content significantly changes the shape of the loop (making it more square-like) and increases coercivity (i.e., from 3 to 24 Oe, for P contents of 8 and 13 at %, respectively).

4. Conclusions

A novel sustainable electrolytic bath has been developed yielding a higher content of P in amorphous smooth, porous-free Fe-W-P alloys. The stable electrolyte consists of an acidic solution with the W source phosphotungstic acid. The Fe-W-P alloy co-deposition was studied by the electrochemical quartz microbalance (EQMB) and EDX analysis. Hydrogen evolution takes place at a negative value of less than −0.4 V (SHE) at a pH of 3 and/or 4, indicating a moderate overpotential vs. the Nernst potential. For a negative potential of −0.85 V co-deposition of the alloy components was evident from a mass and current increase. The overpotential of the W deposition was low (~−0.05 V) in contrast to that during the Fe-P alloy deposition. A reduction of pH leads to an increase in P and W. At higher H^+ concentration, the solubility of $H_3PO_{2(aq)}$ and $H_3PO_{4(aq)}$ species are higher and an increase in deposition rate is observed. The current efficiency increases at pH 4 and is mostly unaffected by the overall current density, due to lower hydrogen evolution that competes with the alloy deposition. The layer morphology was controlled by the competition between hydrogen bubble evolution, their detachment, and the layer growth. The partial current j_e distribution of the corresponding element, i.e., Fe, W, and P, were determined at pH 3 from the EDX data and the weighed coating mass. The current efficiency of alloy electrodeposition increases with current density and, as the partial current of elements. The partial current of the hydrogen evolution is less affected by the current density than all other alloy components. The XRD results show that the Fe-W-P alloys are characterized by an amorphous structure. The alloys exhibit relatively low ferromagnetic signals between M_s = 20 and 40 emu g^{-1} of all measured coatings, possibly due to a mixture of non-ferromagnetic and weakly ferromagnetic phases. The coercively values are typical for a soft magnetic material.

Author Contributions: Conceptualization: N.K., W.H. and W.K.; Methodology: N.K. and W.K.; Validation: N.K., A.M., U.K., J.S., W.H., G.B. and W.K.; Formal Analysis: N.K. and W.K.; Investigation: N.K. and W.K.; Resources: W.H.; Writing—Original Draft Preparation: N.K.; Writing—Review and Editing: N.K., W.K., J.S. and U.K.; Visualization: N.K. and W.K.; Supervision: W.K.; Project

Administration: W.H.; Funding Acquisition: W.H. All authors have read and agreed to the published version of the manuscript.

Funding: The authors acknowledge funding by the HORIZON2020 SELECTA project (No. 642642), the Spanish Government (PID2020-116844RB-C21 and associated FEDER), and the Generalitat de Catalunya (2017-SGR-292).

Institutional Review Board Statement: Not applicable.

Informed Consent Statement: Not applicable.

Data Availability Statement: The data presented in this study are available on request from the corresponding author.

Acknowledgments: Experimental assistance by V. Gman (RENA Technologies Austria GmbH), G. Quorri (RENA Technologies Austria GmbH), and R. Mann (RENA Technologies Austria GmbH) are gratefully acknowledged.

Conflicts of Interest: The authors declare no conflict of interest. The funders had no role in the design of the study; in the collection, analyses, or interpretation of data; in the writing of the manuscript; or in the decision to publish the results.

References

1. Jafari, S.; Beitollahi, A.; Eftekhariyekta, B.; Kanada, K.; Ohkubo, T.; Herzer, G. Microstructural and magnetic properties study of Fe-P rolled sheet alloys. *J. Magn. Magn. Mater.* **2014**, *358*, 38–43. [CrossRef]
2. Vitkova, S.; Kjuchukova, M.; Raichevski, G. Electrochemical preparation of amorphous Fe-P alloys. *J. Appl. Electrochem.* **1988**, *18*, 673–678. [CrossRef]
3. Zecevic, S.K.; Gojkovic, S.L.; Radmilovic, V. Electrochemically deposited thin films of amorphous Fe–P alloy Part I. Chemical composition and phase structure characterization 1. *J. Electroanal. Chem.* **1998**, *448*, 245–252. [CrossRef]
4. Nicolenco, A.; Tsyntsaru, N.; Cesiulis, H. Fe (III)-based ammonia-free bath for electrodeposition of Fe-W alloys. *J. Electrochem. Soc.* **2017**, *164*, D590–D596. [CrossRef]
5. Cserei, A.; Kuzmann, E. Study of the crystallization kinetics in amorphous Fe83P17 alloy. *J. Radional. Nucl. Chem. Lett.* **1994**, *187*, 33–45. [CrossRef]
6. Tsyntsaru, N.; Bobanova, J.; Ye, X.; Cesiulis, H.; Dikusar, A.; Prosycevas, I.; Celis, J.P. Iron-tungsten alloys electrodeposited under direct current from citrate-ammonia plating baths. *Surf. Coatings Technol.* **2009**, *203*, 3136–3141. [CrossRef]
7. Kamei, K.; Maehara, Y. Magnetic properties and microstructure of electrodeposited Fe-P amorphous alloy. *J. Appl. Electrochem.* **1996**, *26*, 529–535. [CrossRef]
8. KHüller; Dietz, G.; Hausmann, R.; Kölpin, K. The composition dependence of magnetization and curie temperature of Fe-P, Co-P and Ni-P. *J. Magn. Magn. Mater.* **1985**, *53*, 103–110. [CrossRef]
9. Kovalska, N.; Pfaffeneder-Kmen, M.; Tsyntsaru, N.; Mann, R.; Cesiulis, H.; Hansal, W.; Kautek, W. The role of glycine in the iron-phosphorous alloy electrodeposition. *Electrochim. Acta* **2019**, *309*, 450–459. [CrossRef]
10. Kovalska, N.; Hansal, W.E.G.; Tsyntsaru, N.; Cesiulis, H.; Gebert, A.; Kautek, W. Electrodeposition and corrosion behaviour of nanocrystalline Fe–P coatings. *Trans. IMF* **2019**, *97*, 89–94. [CrossRef]
11. Kovalska, N.; Tsyntsaru, N.; Cesiulis, H.; Gebert, A.; Fornell, J.; Pellicer, E.; Sort, J.; Hansal, W.; Kautek, W. Electrodeposition of nanocrystalline Fe-P coatings: Influence of bath temperature and glycine concentration on their structural properties, mechanical and corrosion behaviour. *Coatings* **2019**, *9*, 189. [CrossRef]
12. Vernickaite, E.; Antar, Z.Z.; Nicolenco, A.; Kreivaitis, R.; Tsyntsaru, N.; Cesiulis, H. Tribological and Corrosion Properties of Iron-Based Alloys. In Proceedings of the 8th International Science Conference "BALTTRIB 2015", Kaunas, Lithuania, 26–27 November 2015. [CrossRef]
13. Cesiulis, H.; Xie, X.; Podlaha-Murphy, E. Electrodeposition of Co-W alloys with P and Ni. *Medziagotyra* **2009**, *15*, 115–122.
14. Ahmad, J.; Asami, K.; Takeuchi, A.; Louzguine, D.V.; Inoue, A. High strength Ni-Fe-W and Ni-Fe-W-P Alloys Produced by electrodeposition. *Mater. Trans.* **2003**, *44*, 1942–1947. [CrossRef]
15. Zouch, F.; Antar, Z.; Bahri, A.; Elleuch, K.; Ürgen, M. Tribological Study of Fe–W–P Electrodeposited Coating on 316 L Stainless Steel. *J. Tribol.* **2017**, *140*, 011301. [CrossRef]
16. Ahmad, J.; Asami, K.; Takeuchi, A.; Inoue, A. Effect of sodium hypophosphite on the structure and properties of electrodeposited Ni-W-P alloys. *Mater. Trans.* **2003**, *44*, 705–708. [CrossRef]
17. Barbano, E.P.; Carlos, I.A.; Vallés, E. Electrochemical synthesis of Fe-W and Fe-W-P magnetic amorphous films and Fe-W nanowires. *Surf. Coat. Technol.* **2017**, *324*, 80–84. [CrossRef]
18. Thangaraj, N.; Tamilarasan, K.; Sasikumar, D. Effect of phosphorous acid and urea on the ferrous tungsten phosphorous magnetic thin film. *Int. J. Chem. Tech. Res.* **2014**, *6*, 384–390.

19. Mulone, A.; Nicolenco, A.; Fornell, J.; Pellicer, E.; Tsyntsaru, N.; Cesiulis, H.; Sort, J.; Klement, U. Enhanced mechanical properties and microstructural modifications in electrodeposited Fe-W alloys through controlled heat treatments. *Surf. Coat. Technol.* **2018**, *350*, 20–30. [CrossRef]
20. Mulone, A.; Nicolenco, A.; Hoffmann, V.; Klement, U.; Tsyntsaru, N.; Cesiulis, H. In-depth characterization of as-deposited and annealed Fe-W coatings electrodeposited from glycolate-citrate plating bath. *Electrochim. Acta* **2018**, *261*, 167–177. [CrossRef]
21. Brooman, E.W. Wear behavior of environmentally acceptable alternatives to chromium coatings: Cobalt-based and other coatings. *Met. Finish.* **2004**, *102*, 42–54. [CrossRef]
22. Donten, M.X.; Cesiulis, H.; Stojek, Z. Electrodeposition and properties of Ni-W, Fe-W and Fe-Ni-W amorphous alloys. A comparative study. *Electrochim. Acta* **2000**, *45*, 3389–3396. [CrossRef]
23. Sawyer, D.T.; Sobkowiak, A. *Electrochemistry for Chemists*, 2nd ed.; Wiley: New York, NY, USA, 1995.
24. Gamburg, Y.D.; Zangari, G. *Theory and Practice of Metal Electrodeposition*; Springer: Berlin/Heidelberg, Germany, 2011; pp. 1–25. [CrossRef]
25. Bard, A.J.; Faulkner, L.R. *Fundamentals and Fundamentals and Applications*; John Wiley & Sons: Hoboken, NJ, USA, 2015.
26. Janik, M.J.; Campbell, K.A.; Bardin, B.B.; Davis, R.J.; Neurock, M. A computational and experimental study of anhydrous phosphotungstic acid and its interaction with water molecules. *Appl. Catal. Gen.* **2003**, *256*, 51–68. [CrossRef]
27. Noshi, M.N. Phosphotungstic Acid Hydrate. In *Encyclopedia of Reagents for Organic Synthethsis*; John Wiley & Sons, Ltd.: Hoboken, NJ, USA, 2013.
28. Lide, D.R. *CRC Handbook of Chemistry and Physics: A Ready-Reference of Chemical and Physical Data*, 85th ed.; National Institute of Standarts and Technology: Gaithersburg, MD, USA, 2004. [CrossRef]
29. Harris, Q.C. *Quantitative Chemical Analysis*, 7th ed.; W. H. Freeman: New York, NY, USA, 1971; pp. 89–142. [CrossRef]
30. Zhu, T.; Chong, M.N.; Chan, E.S. Nanostructured tungsten trioxide thin films synthesized for photoelectrocatalytic water oxidation: A review. *Chem. Sus. Chem.* **2014**, *7*, 2974–2997. [CrossRef] [PubMed]
31. Anik, M.; Osseo-Asare, K. Effect of pH on the Anodic Behavior of Tungsten. *J. Electrochem. Soc.* **2002**, *149*, B224–B233. [CrossRef]
32. Liu, T.; Ji, B.; Wu, Y.; Liu, Z.; Wang, W. Effects of the pH Value on the Electrodeposition of Fe-P Alloy as a Magnetic Film Material. *J. Phys. Chem. C* **2022**, *126*, 15472–15484. [CrossRef]
33. McCafferty, E. *Thermodynamics of Corrosion: Pourbaix Diagrams*, 1st ed.; Springer-Verlag: New York, NY, USA, 2010. [CrossRef]
34. Brito, M.M.; Zacarin, M.G.; Arruda, R.A.; Barbano, E.P.; Carlos, I.A. Investigation of electrochemical deposition of Fe-P alloys on polycrystalline platinum from an alkaline bath. *Thin Solid Films* **2020**, *699*, 137862. [CrossRef]
35. Tsyntsaru, N.; Cesiulis, H.; Pellicer, E.; Celis, J.P.; Sort, J. Structural, magnetic, and mechanical properties of electrodeposited cobalt-tungsten alloys: Intrinsic and extrinsic interdependencies. *Electrochim. Acta* **2013**, *104*, 94–103. [CrossRef]
36. Nicolenco, A.; Tsyntsaru, N.; Fornell, J.; Pellicer, E.; Reklaitis, J.; Baltrunas, D.; Cesiulis, H.; Sort, J. Mapping of magnetic and mechanical properties of Fe-W alloys electrodeposited from Fe(III)-based glycolate-citrate bath. *Mater. Des.* **2018**, *139*, 429–438. [CrossRef]
37. Tsyntsaru, N.; Cesiulis, H.; Donten, M.; Sort, J.; Pellicer, E.; Podlaha-Murphy, E.J. Modern trends in tungsten alloys electrodeposition with iron group metals. *Surf. Eng. Appl. Electrochem.* **2012**, *48*, 491–520. [CrossRef]
38. Nogués, J.; Apiñaniz, E.; Sort, J.; Amboage, M.; D'Astuto, M.; Mathon, O.; Puzniak, R.; Fita, I.; Garitaonandia, J.S.; Suriñach, S.; et al. Volume expansion contribution to the magnetism of atomically disordered intermetallic alloys. *Phys. Rev. Condens. Matter Mater. Phys.* **2006**, *74*, 1–5. [CrossRef]
39. Pellicer, E.; Varea, A.; Pané, S.; Nelson, B.J.; Menéndez, E.; Estrader, M.; Suriñach, S.; Baró, M.D.; Nogués, J.; Sort, J. Nanocrystalline electroplated Cu-Ni: Metallic thin films with enhanced mechanical properties and tunable magnetic behavior. *Adv. Funct. Mater.* **2010**, *20*, 983–991. [CrossRef]

coatings

MDPI

Article

The Contribution of the Pore Size of Titanium DC (Direct Current) Sputtered Condensation Polymer Materials to Electromagnetic Interruption and Thermal Properties

Hye-Ree Han

Department of Beauty Art Care, Graduate School of Dongguk University, Seoul 04620, Republic of Korea; luckyherry@daum.net

Abstract: Using special materials has been in the spotlight, along with their multifunctional demands, research on electromagnetic interruption, thermal characteristics, biosignal sensors, secondary batteries, etc. In this study, titanium was sputtered into a condensation polymer material and considered in depth in terms of electromagnetic interruption, thermal properties, infrared blocking, etc. As a result of observing the electromagnetic wave shielding effect, the electromagnetic wavelength value decreased from 168.0 to 42.7 to 64.0 when titanium DC sputtered film samples were placed in front of the electromagnetic wave source. The titanium DC sputtered samples significantly reduced electrical resistance compared to the untreated samples. In addition, the IR transmittances of the titanium sputtered specimens were decreased compared to the untreated specimens. When only the cross-section was treated with titanium sputtering and the titanium surface was directed toward the infrared irradiator, the infrared permeability was 64.3 to 0.0%. After taking an infrared thermal image, ΔH, ΔV, ΔS, ΔY, ΔCr, and ΔCb values were calculated. It is believed that the titanium DC sputtered polyamide materials produced in this study can be used for high-functional protective clothing, sensors by applying electromagnetic interruption, IR blocking, and stealth functions.

Keywords: electromagnetic interruption; DC sputtering; titanium

Citation: Han, H.-R. The Contribution of the Pore Size of Titanium DC (Direct Current) Sputtered Condensation Polymer Materials to Electromagnetic Interruption and Thermal Properties. *Coatings* **2023**, *13*, 1756. https://doi.org/10.3390/coatings13101756

Academic Editor: Gergana Alexieva

Received: 11 September 2023
Revised: 30 September 2023
Accepted: 9 October 2023
Published: 11 October 2023

1. Introduction

Titanium (Ti) has a melting point of 1668 °C, higher than that of iron, and has superior corrosion resistance, low thermal expansion heat conductivity, so it has less deformation, and abundant reserves. In addition, titanium is a highly reactive metal, so it reacts with oxygen in the air and forms a titanium oxide film. Titanium has strength and ductility, and is used in various fields, such as electrolytic electrodes, nuclear power generation service devices, corn rods, artificial bones, surgical tools, seawater desalination devices, swivel parts, shafts, LNG seawater cooler tubes, and various mechanical parts.

There are various studies on laser treatment for mechanical stress, the atomic layer deposition of chlorine-containing titanium-zinc oxide nanofilms, 3D-printed titanium lattice structure, titanium alloys, titanium surface modification, porous titanium, and antibacterial applications [1–12].

In addition, sputtering technologies are an environmentally friendly manufacturing technique that does not produce wastewater. This technology also has the advantage of conveniently coating various metals on samples, so it is applied in various ways to high-tech mechanical parts and semiconductors. There are studies on electrical properties, physical properties, thermal stability, biosignal sensors, and hydrophobic thin films using sputtering technology [13–21].

In addition, various studies on the electrical, magnetic, chemical, physical, and sensing characteristics of composite materials related to electrical conductivity are being conducted [22–28].

Regarding infrared rays, research on infrared image fusion, object-oriented attention, infrared and visible light image fusion, and remote sensing images is also being actively conducted [29–32]. The aim of infrared and visible image fusion techniques is to extract and integrate features from images captured using various sensors using specific algorithms to create complementary images that contain both rich detailed features of visible images and target information of infrared images. The technology has far-reaching implications for economic development, including applications in areas such as defense, military, intelligent transportation, and power grid operations [29]. Lv et al. conducted a comprehensive experiment on two infrared datasets, and the results show the efficiency of our approach. We propose a method of generating captions for infrared images based on object-oriented attention. Our approach includes two models: detector and LSTM. The approach of this study is suitable for deployment on embedded devices because it requires fewer resources and is easy to deploy [30]. Xiaodi Xu and others have important applications for the fusion of infrared and visible images in various engineering fields. However, in the current fusion of infrared and visible images, the texture details of the fused images are unclear, and the infrared targets and texture details are disproportionately displayed, resulting in information loss. In this paper, we propose an improved Generative Adversarial Network (GAN) fusion model for the fusion of infrared and visible images [31]. Zhao et al. propose a CNN-based layer adaptive GCP extraction method for TIR RSI. Specifically, the built feature extraction network consists of a primary module and a layer-adaptive module [32].

In addition, there are many studies that have introduced artificial intelligence, electric conductivity, composite multi-function materials, and composite porous materials. This research aims to study, in depth, infrared function, electrical properties, etc., of the titanium sputter technique. In addition, the titanium sputter technique was used to coat the surface for longer than in previous studies to cover the metal grain layer thickly. Moreover, this study aims to examine the correlation between IR images and infrared transmittance. Based on the results of the research data, the applicability to industrial fields is considered.

In this paper, the electromagnetic wave blocking performance of titanium sputtering-treated polyamide materials, stealth effects on infrared cameras, electrical conductivity, transmittance of IR, electromagnetic wave blocking properties, and thermal properties were examined using various methods. In addition, a titanium sputtered samples were prepared by changing the pore size of the polyamide sample.

Therefore, titanium sputtering was performed by dividing polyamide into films, plain fabrics, and mesh (steps 1 to 5) to observe various characteristics according to sample pore size. Furthermore, based on the study results, electromagnetic blocking performance, heat transfer characteristics, stealth effects on IR cameras, and the applicability of high-functional smart materials were considered.

2. Materials and Methods

The materials used for titanium DC sputtering treatment are polyamide materials (film, fabric, and mesh, steps 1–5). Titanium sputter process was prepared with various pore sizes. In addition, the sample characteristics are as shown in Table 1.

Table 1. Information of base polyamide materials.

	Poly Amide Film	Poly Amide Fabric	Poly Amide Net 1	Poly Amide Net 2	Poly Amide Net 3	Poly Amide Net 4	Poly Amide Net 5
Sample code	TF1	TA1	TN1	TN2	TN3	TN4	TN5
Sample thickness (mm)	0.09	0.15	0.08	0.10	0.15	0.11	0.19
Weave type	Film	Plain weave	Plain mesh	Plain mesh	Plain mesh	Plain mesh	Plain mesh
Pore size (μm^2)	0	1200	3564	35,696.3	94,440.4	136,476.2	294,825.5

The titanium sputtering process applied to the polyamide materials were as represented in Table 2. Sputter coater (SORONA, SRN-120, Pyeongtaek, Korea) is the device used for titanium sputtering treatment. When the titanium sputter technique was performed, the base polyamide material diameter of the circular shape was 19.5 cm. In this study, the pore size was adjusted by changing the density of warp and weft yarns during weaving. Sputtering is a type of vacuum deposition method commonly used in integrated circuit production line processes, which accelerates gas, such as ionized argon, at a relatively low vacuum degree, collides with a target, and ejects atoms to create a film on a substrate such as a wafer or glass.

Table 2. Titanium sputter conditions.

Time (s)	2400
Process pressure (Torr)	6 m
Gas (sccm)	Ar 40
Power (W)	DC 600
Machine	SRN-120

As a result of FE-SEM with a digital microscope after titanium sputter process, it was revealed that titanium was well formed on the polyamide surface in all sputtered specimens.

In the case of electromagnetic wave blocking performance evaluation, an electromagnetic-wave-measuring device (9024-EN-00, Electromagnetic radiation tester, China) was used to place a sample between the electromagnetic wave generator and the electromagnetic-wave-measuring device to measure the degree of electromagnetic wave blocking.

Infrared intensity tester (5 mm Infrared LED, Infrared Emitting Diodes, EVERLIGHT) was utilized for IR transmittance. The IR strength irradiated to the specimens was 200 W/m^2. In addition, for major infrared wavelength, it was 940 nm.

For stealth characteristics, IR Thermographic camera (Fliri7) was used. The specimen was placed close to the heat source and a thermal images were obtained using an IR thermal camera. The hidden effect and values of H, V, S, Y, Cr, and Cb were recorded for infrared thermal imaging cameras utilizing the Color Inspector 3D program. In addition, more quantitative color differences were studied for H, S, V, Y, Cb, and Cr data.

In addition, ΔH, ΔV, ΔS, ΔY, ΔCr, and ΔCb data were calculated. The data of ΔH, ΔS, and ΔV values are expressed in the following Equations (1)–(3).

$$\Delta H = H\_treated - H\_untreated \tag{1}$$

$$\Delta S = S\_treated - S\_untreated \tag{2}$$

$$\Delta V = V\_treated - V\_untreated \tag{3}$$

in the definitions of Equations (1)–(3) are as follows.

H_untreated: data H of untreated samples;
H_treated: data H of sputtered samples;
S_untreated: data S of the untreated samples;
S_treated: data S of the sputtered samples;
V_untreated: data V of the untreated samples;
V_treated: data V of the sputtered samples.

Equation (1) shows the degree of change in the "H" data due to titanium sputtering treatment by subtracting the "H" data of the untreated specimen from the "H" data of the titanium sputtered sample. The value "H" in the concept of color space means a color angle (hue). In the Equation (2), the change in the "S" data due to titanium sputter technique is the data obtained by subtracting the "S" data of the "untreated fabric" from the "S" data of the titanium sputtered sample. The "S" value in the color space represents the concentration and saturation of the color. Equation (3) demonstrates that in this case, the "V" data of the untreated specimen are subtracted from the "V" data of the titanium sputtered sample, indicating the color difference caused by titanium sputtering technique. The "V" value in the color space represents the degree of brightness and value. Using the

"T" data of Equation (7), the color difference between the titanium sputtered specimen and the untreated specimen in color space was calculated.

ΔY, ΔCb, and ΔCr values were calculated as follows.

$$\Delta Y = Y\_treated - Y\_untreated \quad (4)$$

$$\Delta Cb = Cb\_treated - Cb\_untreated \quad (5)$$

$$\Delta Cr = Cr\_treated - Cr\_untreated \quad (6)$$

$$\Delta T = \sqrt{(\Delta Y)^2 + (\Delta Cb)^2 + (\Delta Cr)^2} \quad (7)$$

Definitions of Equations (4)–(7) are provided below.
Y_untreated: data Y of untreated specimen;
Y_treated: data Y of sputtered specimen;
Cb_untreated: data Cb of the untreated specimen;
Cb_treated: data Cb of the sputtered specimen;
Cr_untreated: data Cr of the untreated specimen;
Cr_treated: data Cr of the sputtered specimen.

Equation (4) shows a change in the "Y" data due to titanium sputtering technique by subtracting the "Y" data of the untreated fabric from the "Y" data of the titanium sputtering treatment sample. Equation (5) shows the change in the data of "Cb" due to titanium sputtering technique by subtracting the data of "Cb" of "untreated sample" from the data of "Cb" of the titanium sputtering specimens. Equation (6) shows the change in Cr data by titanium sputter process by subtracting the data "Cr" of the untreated specimens from the data "Cr" of the titanium sputtered specimens.

3. Results and Discussions

3.1. Surface Geometry

FE-SEM photography images were captured to observe the nano-shape formed on the titanium sputtering-treated sample (Figure 1). As a result of observing the micrograph, it was possible to capture that a titanium layer was formed on the surface of all titanium sputtering-treated samples. In addition, that a titanium surface in the form of a meteorite was formed in all samples could be captured. Meteorite-shaped titanium sizes were captured at 45.58–80.98 and 53.63–67.24 nm for TF1 and TA1, respectively. In the case of TN1, TN2, TN3, and TN4, titanium sizes were 83.37 to 117.2, 95.24 to 100.1, 62.19 to 118.0, and 55.15 to 79.60 nm, respectively. In previous studies, after sputtering treatment, FE-SEM image observations showed that nanograins were formed on the surface [33], and in this research, they showed a meteorite shape rather than a grain shape on the surface. Since the sputtering process time was longer than that of previous studies, it was judged that the titanium layer was more coated, resulting in an irregular meteorite shape. Moreover, the result of FE-SEM prompts research in more detail on what the thickness of the titanium layer is on the specimens in Figure 2 is equal to. In the case of TF1 and TA1, the thickness of the titanium layer was captured as 718.5 to 751.4 and 470.8 to 521.7 nm, respectively. In the case of TN1, TN2, TN3, and TN4, the thickness of the titanium layer was 606.0 to 619.7, 553.9 to 578.6, 693 to 803.5, and 551.2 to 567.6 nm, respectively. The EDX data are as shown in Figure 3, which proved that titanium materials were well coated on the sample surface.

3.2. Characteristics of Electromagnetic Interference

The electronic wave blocking performance of the titanium sputtered specimens is considered to be related to the pore size. As a result of observing the electromagnetic wave shielding effect, the electromagnetic wavelength (V/m) value decreased from 168.0 to 42.7 to 64.0 when titanium sputtering-treated film samples were placed in front of the electromagnetic wave source (Figure 4). In addition, the titanium coating layer was measured to block electromagnetic waves better when it was facing away from electromagnetic irradiator than when facing the electromagnetic irradiator (titanium phase down, 42.7 V/m). When titanium sputtering was performed on other polyamide mesh, it was confirmed that there was little electromagnetic wave blocking effect. The smaller the pore size (294,825.5 → 0 μm^2), the better the electromagnetic wave blocking performance. In previous studies with aluminum sputtered specimens, as the pore size decreased, the electromagnetic wave transmission values were from 25.7 mG to 49.0 mG, and the electromagnetic wave blocking performance decreased [33]. It is believed that electromagnetic waves pass between the pores using the pores between the samples. However, in the case of films, it is thought that the titanium layer blocks electromagnetic waves because there are no pores. However, in order to more completely block electromagnetic waves so that the V/m value becomes 0, it is believed that the titanium layer must be sputtered thicker on the film.

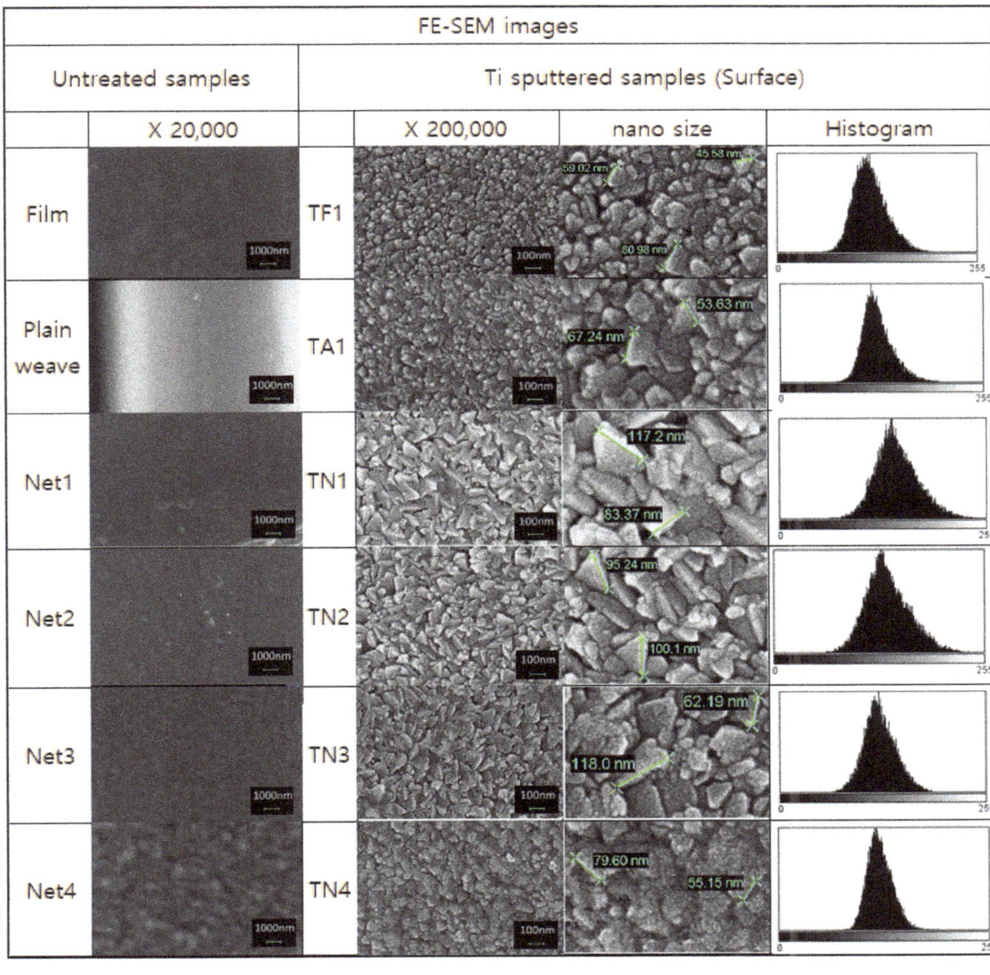

Figure 1. Surface characteristics of titanium DC sputtered polyamide materials.

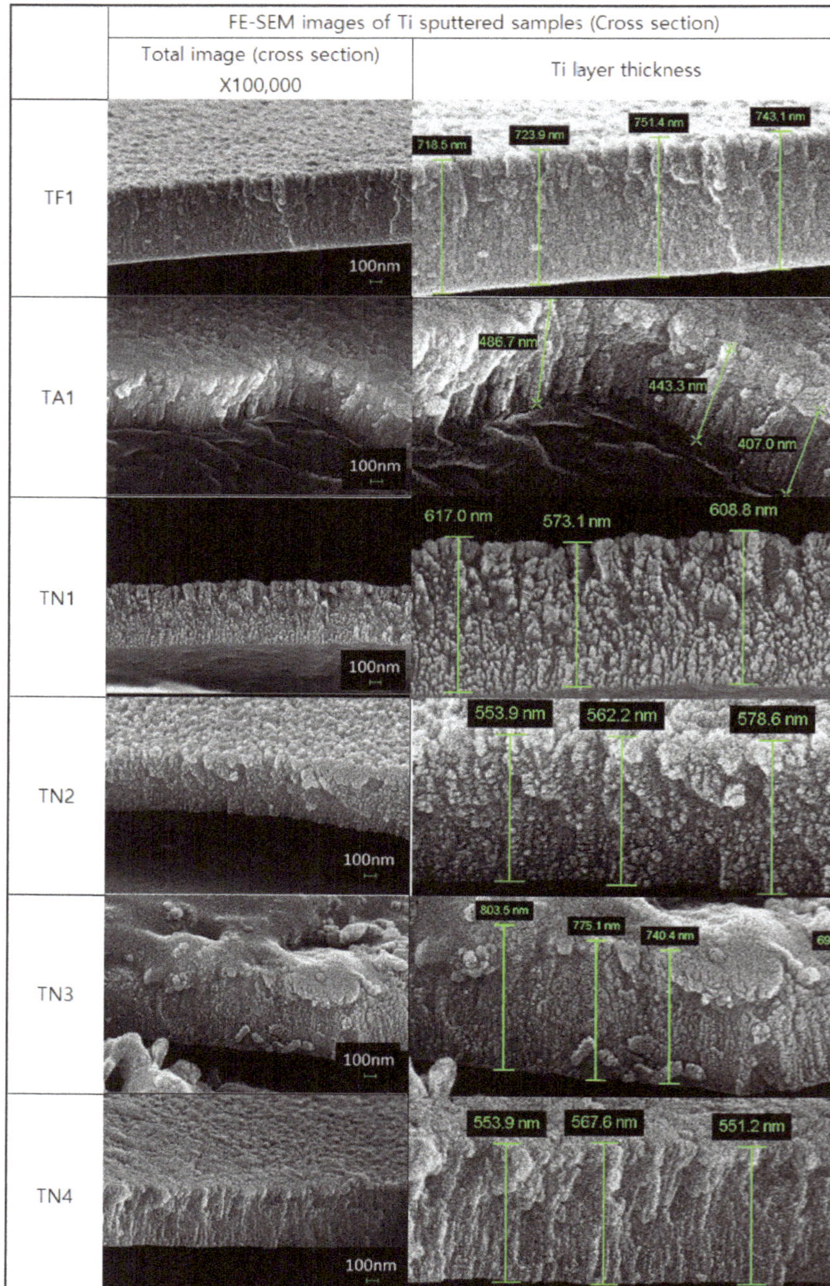

Figure 2. Cross-section characteristics of titanium DC sputtered polyamide materials.

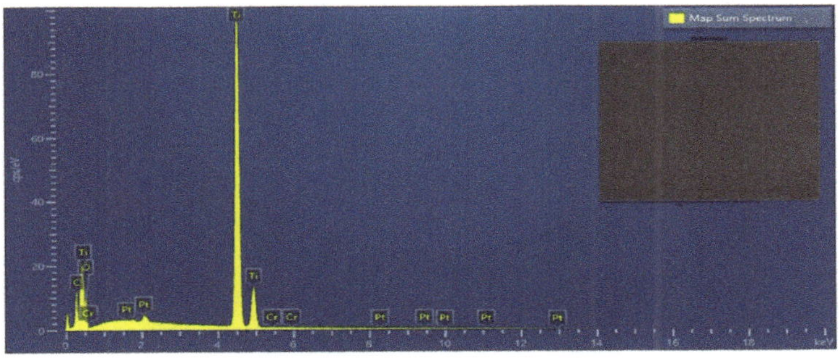

Element	Line Type	Apparent Concentration	k Ratio	Wt%	Wt% Sigma	Atomic %	Standard Label	Factory Standard
C	K series	5.90	0.05899	6.69	0.15	17.77	C Vit	Yes
O	K series	9.98	0.03358	15.09	0.32	30.11	SiO2	Yes
Ti	K series	154.31	1.54309	78.17	0.32	52.09	Ti	Yes
Cr	K series	0.09	0.00091	0.05	0.06	0.03	Cr	Yes
Total:				100.00		100.00		

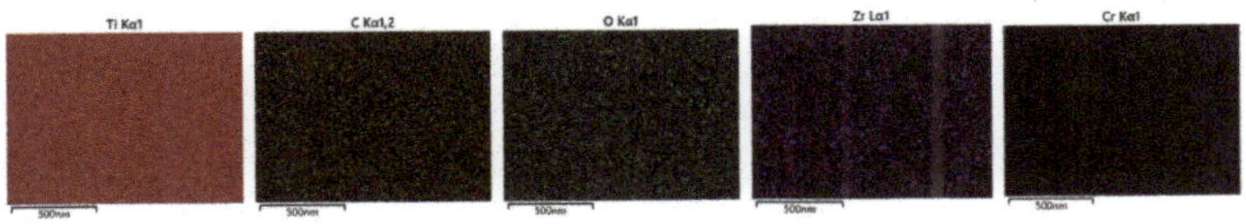

(**a**)

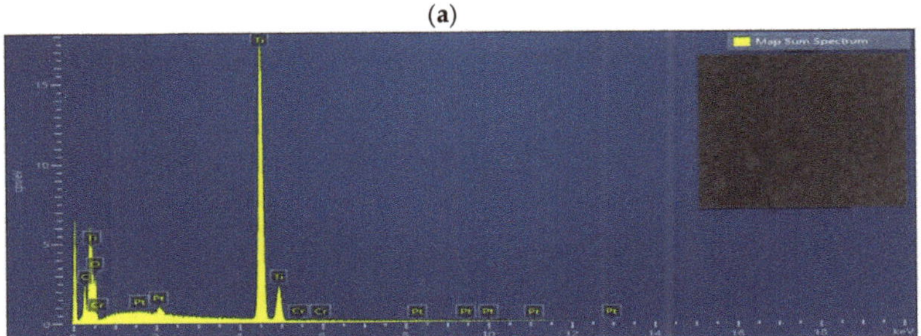

Element	Line Type	Apparent Concentration	k Ratio	Wt%	Wt% Sigma	Atomic %	Standard Label	Factory Standard
C	K series	5.06	0.05056	4.59	0.25	12.56	C Vit	Yes
O	K series	14.71	0.04950	15.98	0.58	32.87	SiO2	Yes
Ti	K series	175.79	1.75786	79.43	0.59	54.57	Ti	Yes
Cr	K series	0.00	0.00000	0.00	0.16	0.00	Cr	Yes
Total:				100.00		100.00		

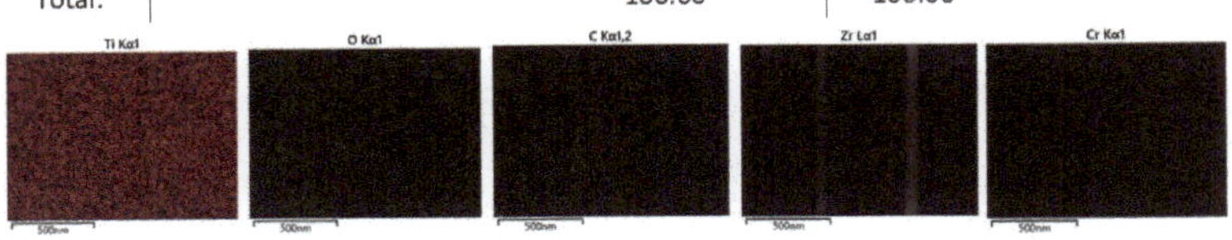

(**b**)

Figure 3. *Cont.*

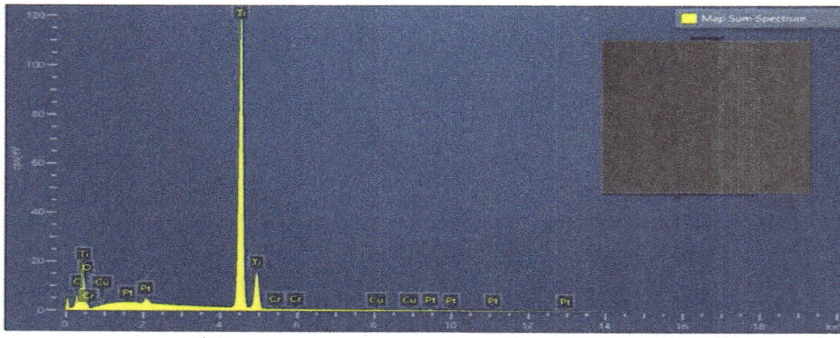

Element	Line Type	Apparent Concentration	k Ratio	Wt%	Wt% Sigma	Atomic %	Standard Label	Factory Standard
C	K series	2.02	0.02015	2.41	0.11	7.30	C Vit	Yes
O	K series	7.57	0.02548	12.28	0.28	27.93	SiO2	Yes
Ti	K series	167.34	1.67342	85.25	0.30	64.74	Ti	Yes
Cr	K series	0.00	0.00000	0.00	0.06	0.00	Cr	Yes
Cu	L series	0.04	0.00040	0.06	0.10	0.03	Cu	Yes
Total:				100.00		100.00		

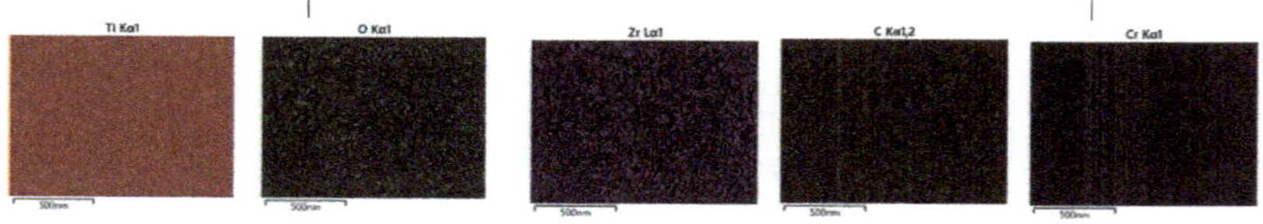

(**c**)

Figure 3. EDX results of molybdenum sputtered polyamide specimens: (**a**) TF1, (**b**) TA1, and (**c**) TN1.

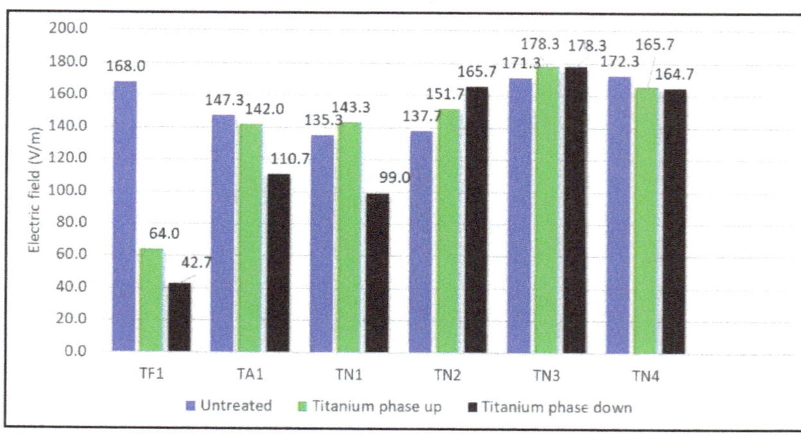

(**a**)

Figure 4. *Cont.*

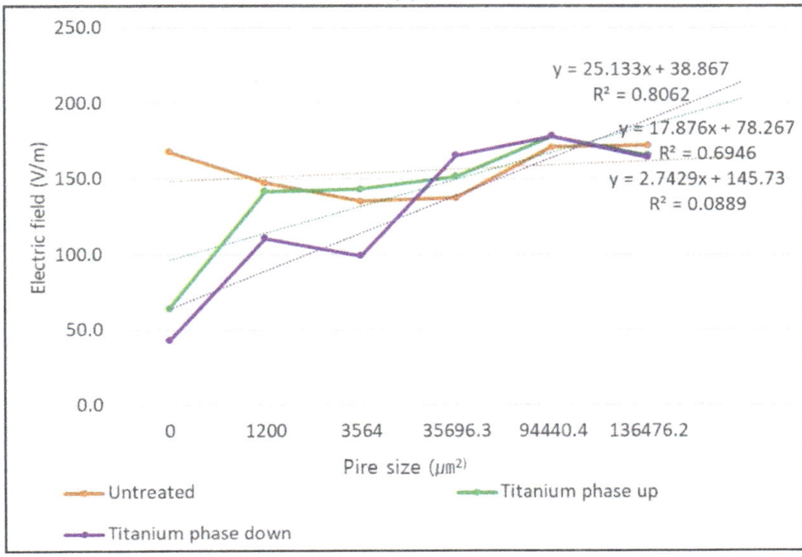

(b)

Figure 4. Characteristics of electric field of specimens. (**a**) Electromagnetic interruption of titanium DC sputtered polyamide specimens. (**b**) Regression line of electric field.

3.3. Characteristics of Electrical Conductivity

The results of the electrical conductivity of titanium sputtered specimens are considered to be related to structure and pore size. Compared to the untreated specimens, the electrical resistance data of titanium sputtered film and mesh specimen were decreased (Figure 5). In the results of the untreated specimens, the electrical resistance of TF1, TA1, and TN1–5 all showed "over load". However, after the titanium sputtering process, the electrical resistance values of all specimens other than NFa were significantly reduced. In this research, the sputtering time of the titanium sputtering-treated TN1 to TN5 samples is long, so the thickness of the titanium coating layer is thick, and the current is not cut off and electrical conductivity is reduced. However, in the case of the TA1 sample, the electrical conductivity was relatively high, even when the metal sputtering process was performed under the same conditions. This is believed to be due to the dense intersection of weave yarn, which has many deadlock points, and the height of the weave yarn furrow is higher than that of the mesh. Previous studies using aluminum sputtering samples showed a similar trend: films showed good electrical conductivity, while fabrics did not. It was explained that this is judged to be due to the fact that the metal layer is cut off due to the fraying of the warp and weft yarns, and the lack of current flow [33].

3.4. Infrared Transmittance Behavior

In this paper, infrared transmittance experiments of untreated samples and cross-section titanium sputtered samples were conducted. The experimental data are represented in Figure 6. An infrared irradiator was placed on the left, and infrared measuring apparatus was placed on the right. In addition, the titanium sputtered specimens were placed between the irradiator equipment and the measuring apparatus. As a result of the experiment, the sample treated with titanium sputter treatment showed a significant reduction in IR transmission data.

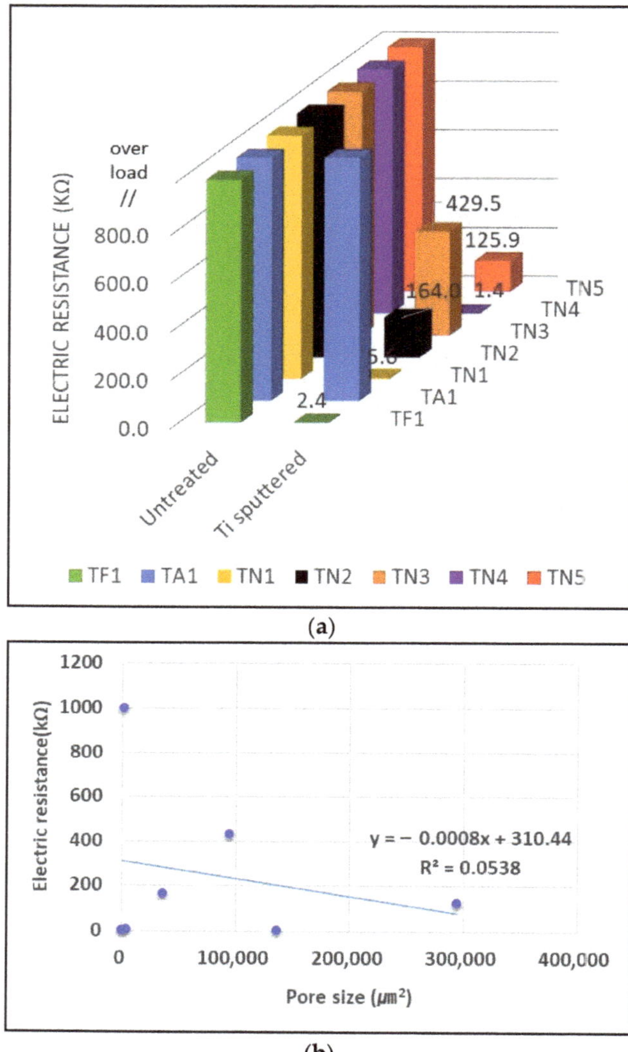

Figure 5. Characteristics of electrical conductivity of specimens. (**a**) Electrical resistance of specimens. (**b**) Correlation between electrical resistance and pore size.

In addition, when titanium sputtering was performed on the cross-section and the titanium surface faced the infrared measuring apparatus (titanium phase: back), the infrared transmittance was 67.2%–0.0%. Namely, the changes in the infrared transmission data according to the direction of the titanium sputtered layer were not large. However, However, when the titanium surface was directed to the infrared irradiator, the IR transmittance data was slightly reduced compared to when the titanium surface was directed to the infrared measuring apparatus. In this research, since the thickness of the titanium layer was made thicker than in past studies, it is judged that the difference in infrared transmittance (%) according to the direction of the titanium surface was very small.

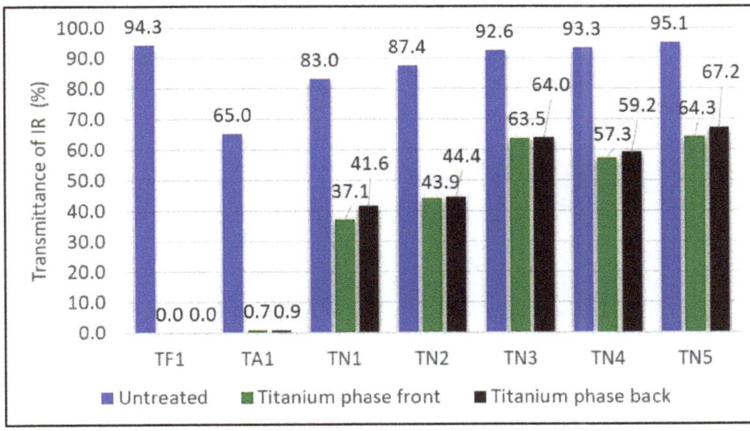

(a)

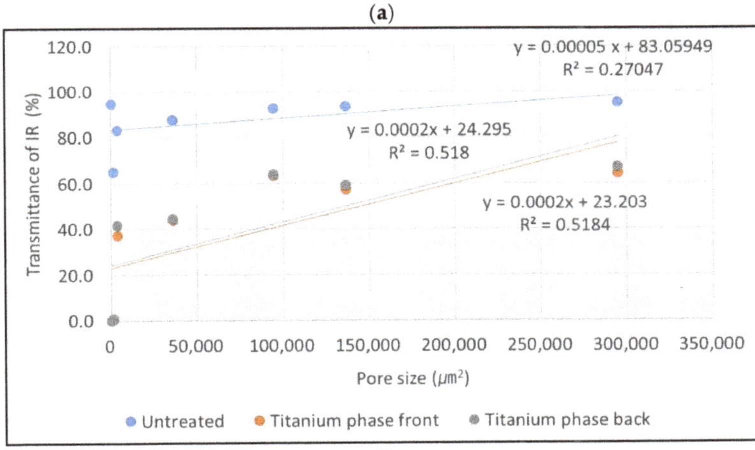

(b)

Figure 6. Transmittance of IR for each specimen. (**a**) IR transmittance. (**b**) Correlation between IR transmittance and pore size.

3.5. Infrared Image Stealth Characteristics According to Heat Transfer at High Temperature

In this research, thermal characteristics attributed to high degree heating sources were examined using IR thermal cameras (Figures 7–10). Photographing was performed with an IR camera with a distance of 0 cm between the sample and the heat source. Only the cross-section was photographed by changing the direction of the titanium sputtering sample. The outward temperature of the heating source was 45.1 to 49.0 °C.

When the titanium layer was facing towards to outside air, the surface temperature was much less than the heating source. When the titanium layer was facing the outside air, the surface temperature was 32.5 °C, and when the titanium layer was directed towards the heating source, the surface temperature was 38.5 °C. However, when the titanium layer faced the heating source, the heating source temperature appeared in the IR image as it was, and there was little hidden effect.

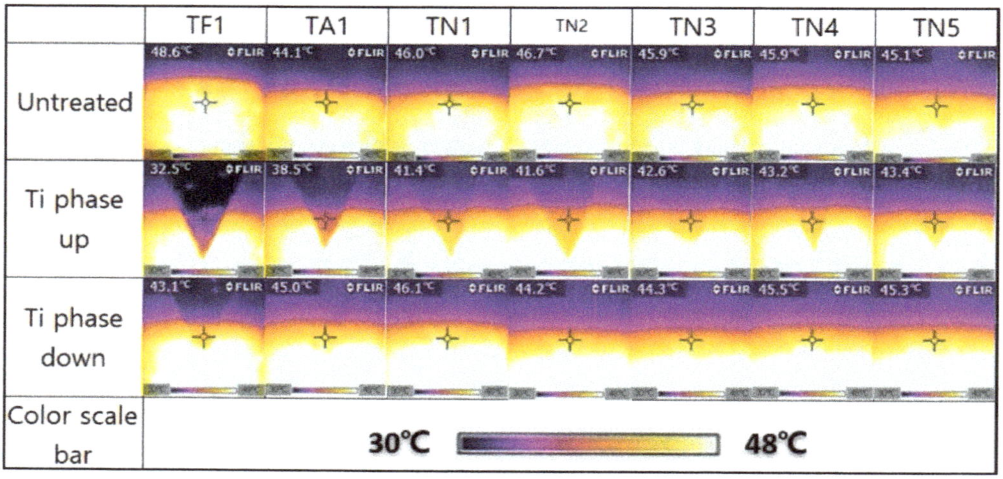

Figure 7. IR thermal images according to different densities.

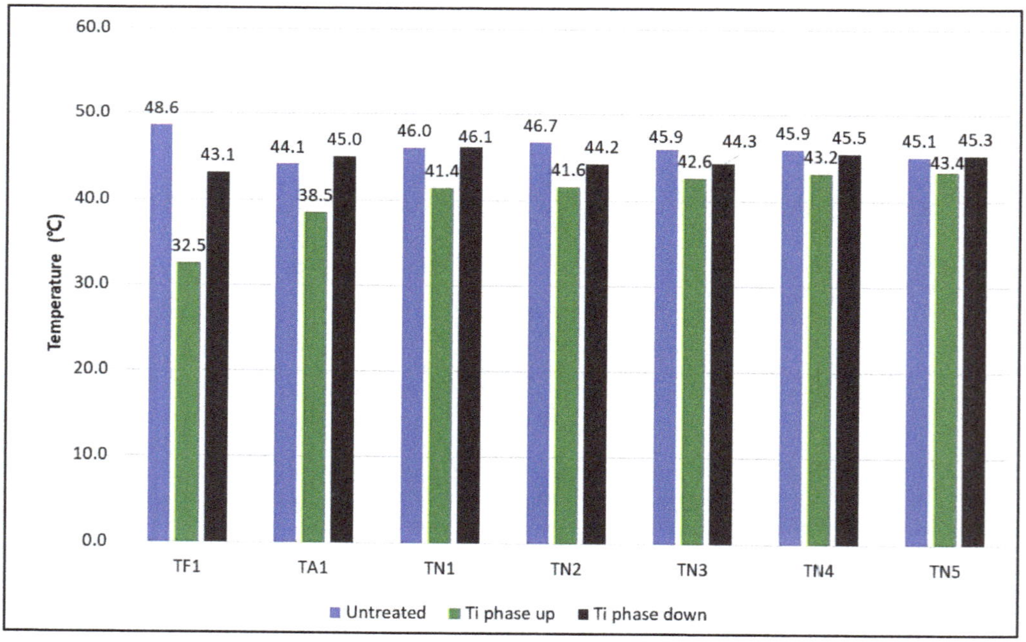

Figure 8. Surface temperature with different specimens (distance between heat source and sample: 0 mm).

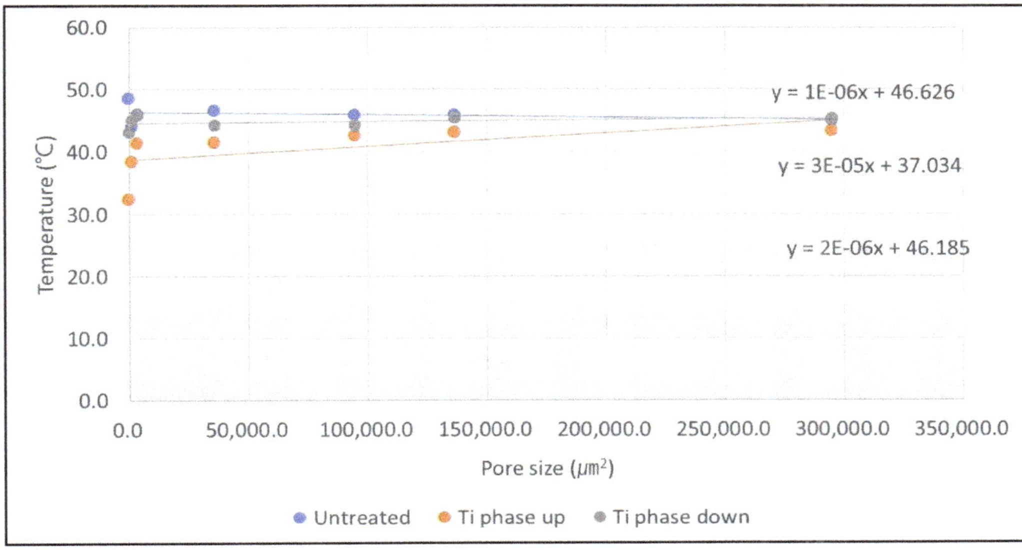

Figure 9. Correlation between pore size and surface temperature.

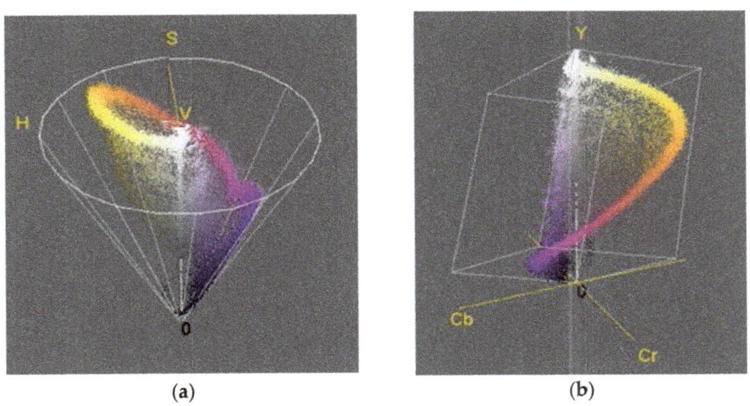

Figure 10. Three-dimensional color distribution. (**a**) Schematic illustration of H, S, and V. (**b**) Schematic illustration of Y, Cb, and Cr.

Moreover, in the case of the mesh specimen, when the titanium sputtering layer was facing the outside air and when the mesh's pore size was larger (3564 → 294,825.5), the surface temperature was from 41.4 to 43.4 °C. When the pore size was bigger, the heat of the heating source was emitted to the outside as it was, and it is believed that the surface temperature was the same as the heating source temperature. In prior studies using aluminum sputtering treatment materials, when the metal layer was directed to the outside air, the surface temperature tended to increase from 27.6 to 34.5 °C as the pore size increased [33].

Furthermore, when the titanium layer of the titanium sputtered mesh sample faces a person's body, the surface temperature was 43.1 to 46.1 °C, indicating a high surface temperature close to the heating source.

These results are judged to be due to the rapid release of heat from the heat source by the titanium layer into the outside air.

After conducting IR imaging, H, S, and V data (Table 3) were obtained using a program (Color Inspector 3D, Image J), and ΔH, ΔV, ΔS, and ΔE data were calculated (Table 4). The measurement Images were shown in Figure 7. The lower right corner of the cross is expressed in Figure 7. And correlation between ΔE and pore size were shown in Figure 11. In addition, the data of ΔH, ΔS, ΔV, and ΔE values are the expressions "Equations (1)–(3)". The H, S, and V data of the outside air were "269, 49, and 53", respectively. Moreover, the H, S, and V data of the heating source were "52, 67, and 93", respectively.

Table 3. H, V, and S data of untreated and titanium sputtered polyamide materials.

	Untreated			Titanium Phase: Up			Titanium Phase: Down		
	H	**S**	**V**	**H**	**S**	**V**	**H**	**S**	**V**
TF1	72	12	92	280	55	61	57	49	87
TA1	47	69	93	345	51	79	52	58	91
TN1	60	28	89	37	93	96	62	34	89
TN2	60	27	92	39	82	89	55	59	91
TN3	60	35	90	44	81	92	52	73	92
TN4	56	38	91	45	85	93	62	27	90
TN5	49	69	92	47	84	91	57	48	90

Table 4. ΔH, ΔV, and ΔS results based on infrared thermal image.

	Titanium Phase: Up				Titanium Phase: Down			
	ΔH	**ΔS**	**ΔV**	**ΔE**	**ΔH**	**ΔS**	**ΔV**	**ΔE**
TF1	208	43	−31	214.6	−15	37	−5	40.2
TA1	298	−18	−14	298.9	5	−11	−2	12.2
TN1	−23	65	7	69.3	2	6	0	6.3
TN2	−21	55	−3	58.9	−5	32	−1	32.4
TN3	−16	46	2	48.7	−8	38	2	38.9
TN4	−11	47	2	48.3	6	−11	−1	12.6
TN5	−2	15	−1	15.2	8	−21	−2	22.6

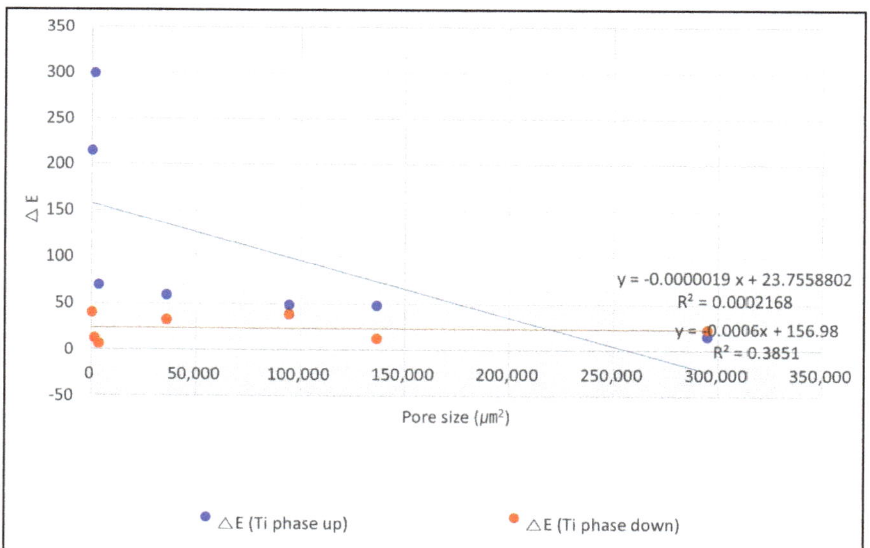

Figure 11. Correlation between ΔE and pore size.

The H, S, and V values of "untreated samples" and "Titanium phase down" (when the titanium surface of the cross-sectional sputtered specimen faces the people body) were relatively similar in all specimens. However, the difference in pore size was not significant. H values of all untreated samples were 47 to 72, S values were 12 to 69, and V values were 89 to 92. For all samples of titanium phase down, H values were 52–62, S values were 27–73, and V values were 87–92. It shows a similar aspect to the image of Figure 7. The small absolute value (6.3~40.2) of the specimens of titanium phase down indicates that the difference in H, S, and V data between titanium phase down samples and untreated samples is small. Moreover, in this case, the hidden effect on IR thermal imaging is small. It shows a similar aspect to the thermal image of Figure 7.

On the other hand, the HSV data of "Titanium phase up" (when the titanium surface of the cross-sectional sputtered specimen faces the outside air) represented a different tendency from "untreated sample" and "Titanium phase down". For the data of all specimens in the titanium phase up, the H data were 37–345, the S data were 51–93, and the V data were 61–96. In the H data, the H data tended to decrease as pore size increased. In the case of the V data, they showed a tendency to increase as the pore size of the sample increased. The large amount of absolute data (15.2~298.9) of ΔE of titanium phase up indicates that the difference in H, S, and V data between titanium phase up samples and untreated samples is large. It can be seen that the titanium phase up sample indicates that there is an alternative hidden effect on the IR imaging.

The Y, Cb, and Cr data (Table 5) of "untreated samples" and "Titanium phase down" were very similar in all samples, regardless of pore size. However, the difference in pore size was not significant. The Y values of all untreated samples were 194–231, Cb data were −13–83, and Cr data were −4–29. For all samples of titanium phase down, Y data were 199–221, Cb data were −42–76, and Cr data were 7–25. The fact that the absolute data of ΔY, ΔCb, and ΔCr of the specimens of titanium phase down are 5~29, 7~63, and 3~29, respectively, indicates that the difference in Y, Cb, and Cr data between the titanium phase down specimen and the untreated specimen is small, and explains the hidden effect on the IR imaging.

Table 5. Y, Cb, and Cr data of untreated and titanium sputter treated specimens.

	Untreated			Titanium Phase: Up			Titanium Phase: Down		
	Y	Cb	Cr	Y	Cb	Cr	Y	Cb	Cr
TF1	228	−13	−4	55	52	35	199	−76	25
TA1	194	−83	29	113	−11	64	204	−68	20
TN1	217	−28	2	166	−89	57	212	−45	7
TN2	231	−32	7	176	−90	49	204	−72	19
TN3	223	−35	3	186	−94	37	201	−72	17
TN4	212	−49	12	181	−88	39	221	−42	9
TN5	202	−76	27	192	−90	32	216	−59	12

In the case of the Y value, the pore size of the specimen increased as it increased. The absolute data (Table 6) of ΔY, ΔCb, and ΔCr in the titanium phase up were 10~173, 14~72, and 5~55, respectively. In the case of ΔT (Figure 12), when pore size was decreased, the larger the amount of absolute data appeared. The large absolute data of ΔY, ΔCb, and ΔCr indicate that the difference in Y, Cb, and Cr data between titanium phase up samples and untreated specimens are large. This indicates that dense titanium phase-up specimens have an alternative hidden effect on IR imaging.

Table 6. ΔY, ΔCr, ΔCb, and ΔT data based on IR thermal images.

	Titanium Phase: Up				Titanium Phase: Down			
	ΔY	ΔCb	ΔCr	ΔT	ΔY	ΔCb	ΔCr	ΔT
TF1	−173	65	39	188.9	−29	−63	29	75.2
TA1	−81	72	35	113.9	10	15	−9	20.1
TN1	−51	−61	55	96.7	−5	−17	5	18.4
TN2	−55	−58	42	90.3	−27	−40	12	49.7
TN3	−37	−59	34	77.5	−22	−37	14	45.3
TN4	−31	−39	27	56.7	9	7	−3	11.8
TN5	−10	−14	5	17.9	14	17	−15	26.6

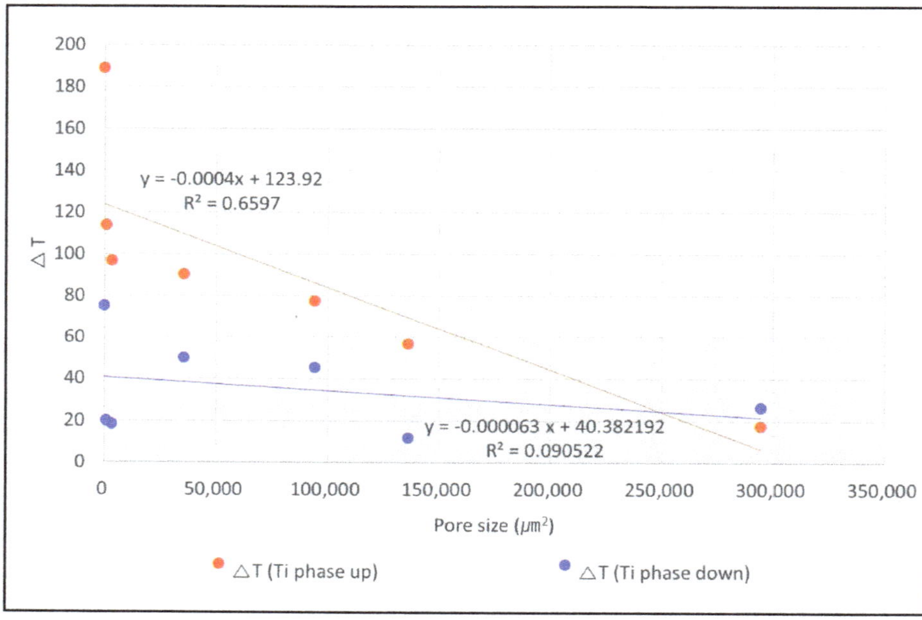

Figure 12. Correlation between ΔT and pore size.

4. Conclusions

In this study, titanium sputtered specimens were closely observed to determine their electromagnetic blocking properties, hidden effect for IR cameras, electrical characteristics, and thermal properties according to poresize (Table 7). Polyamide specimens with different pore sizes were prepared (film, fabric, and mesh) as base materials for titanium sputtering treatment. Moreover, stealth function for IR thermal imaging was evaluated with H, S, V, Y, Cb, Cr, etc.

As a result of the electrical resistance experiment to observe the electrical conductivity, all untreated specimens had a high amount of resistance data to the extent of overloading, and no electrical conductivity was shown. However, titanium sputtered samples had significantly less electrical resistance data. As a result of infrared transmittance experiments, the infrared transmittance of the titanium sputtered specimens was significantly reduced compared to the untreated specimens. In the results of untreated samples, the transmittance was 95.1 to 65.0%. When the titanium surface faced the IR irradiator and pore size was decreased, the IR transmittance was decreased (64.3 to 0.0%). In addition, when the titanium surface faced the IR measuring device and pore size was decreased, the IR transmittance was decreased (67.2 to 0.0%). That is, the change in the IR transmission data according to the direction of the titanium sputtered layer was not large. When the pore size of the specimen is larger, the transmittance is higher. In the IR thermal imaging, in the case of the mesh specimen, when the titanium sputtering layer faced the outside air and the mesh's pore size was larger (TN1 to TN5), the surface temperature was 41.4 to 43.4 °C, and hidden characteristics were decreased. As the pore size of the mesh increases, the heat of the human body escapes to the outside air, and the surface temperature is believed to be the same as the heating source temperature. After taking IR thermal imaging, the data of H, S, V, Y, Cb, and Cr were measured to confirm the stealth effect on the quantitative infrared camera, and the values of ΔH, ΔV, ΔS, ΔY, ΔCr, and ΔCb were calculated. The data of H, V, and S of the "untreated sample" and "Titanium phase down" were very similar in all samples, and the difference according to pore size was not significant. In this research, the results indicate that dense titanium phase up samples have an alternative hidden effect on IR cameras. This fact is interpreted as having little influence on the infrared transmittance of the infrared thermal image, and changes in wearing direction, multi-layer manufacturing depending on the purpose of wearing can be considered for compatibility depending on external environmental factors.

Table 7. Comparative data according to pore size of sputtered specimens.

Pore Size of Sputtered Specimens (μm^2)	Electromagnetic Field of Titanium Phase Up (V/m)	Electrical Resistance (kΩ)	IR Transmittance of Titanium Phase Front (%)	Surface Temperature of Titanium Phase Up (°C)	ΔE Value of Titanium Phase Up	ΔT Value of Titanium Phase Up
0	64.0	2.4	0	32.5	214.6	188.9
1200	142.0	1000.0	0.7	38.5	298.9	113.9
3564	143.3	5.6	37.1	41.4	69.3	96.7
35,696	151.7	164.0	43.9	41.6	58.9	90.3
94,440	178.3	429.5	63.5	42.6	48.7	77.5
136,476	165.7	1.4	57.3	43.2	48.3	56.7

The titanium sputtering polyamide samples developed in this study showed excellent electromagnetic wave blocking performance, stealth function for infrared thermal imaging detectors, and IR-blocking characteristics. These findings are expected to be applicable to high-functional clothing, sensors, stage costumes, batteries, and more.

Funding: This research received no external funding.

Institutional Review Board Statement: Not applicable.

Informed Consent Statement: Not applicable.

Data Availability Statement: Not applicable.

Conflicts of Interest: The author declares no conflict of interest.

References

1. Chen, H.; Feng, R.; Xia, T.; Wen, Z.; Li, Q.; Qiu, X.; Huang, B.; Li, Y. Progress in Surface Modification of Titanium Implants by Hydrogel Coatings. *Gels* **2023**, *9*, 423. [CrossRef] [PubMed]
2. Chumaevskii, A.; Amirov, A.; Ivanov, A.; Rubtsov, V.; Kolubaev, E. Friction Stir Welding/Processing of Various Metals with Working Tools of Different Materials and Its Peculiarities for Titanium Alloys: A Review. *Metals* **2023**, *13*, 970. [CrossRef]
3. Guo, A.X.Y.; Cao, B.; Wang, Z.; Ma, X.; Cao, S.C. Fabricated High-Strength, Low-Elastic Modulus Biomedical Ti-24Nb-4Zr-8Sn Alloy via Powder Metallurgy. *Materials* **2023**, *16*, 3845. [CrossRef] [PubMed]
4. Jażdżewska, M.; Bartmański, M.; Zieliński, A.; Kwidzińska, D.B. Effect of Laser Treatment on Intrinsic Mechanical Stresses in Titanium and Some of Its Alloys. *Appl. Sci.* **2023**, *13*, 6276. [CrossRef]
5. Kovács, Á.É.; Csernátony, Z.; Csámer, L.; Méhes, G.; Szabó, D.; Veres, M.; Braun, M.; Harangi, B.; Serbán, N.; Zhang, L.; et al. Comparative Analysis of Bone Ingrowth in 3D-Printed Titanium Lattice Structures with Different Patterns. *Materials* **2023**, *16*, 3861. [CrossRef] [PubMed]

6. Lu, X.; Shu, C.; Zheng, Z.; Shu, X.; Chen, S.; Essa, K.; Li, Z.; Xu, H. Effects of L-PBF Scanning Strategy and Sloping Angle on the Process Properties of TC11 Titanium Alloy. *Metals* **2023**, *13*, 983. [CrossRef]
7. Nazarov, D.; Kozlova, L.; Rudakova, A.; Zemtsova, E.; Yudintceva, N.; Ovcharenko, E.; Koroleva, A.; Kasatkin, I.; Kraeva, L.; Rogacheva, E.; et al. Atomic Layer Deposition of Chlorine Containing Titanium–Zinc Oxide Nanofilms Using the Supercycle Approach. *Coatings* **2023**, *13*, 960.
8. Pietzka, S.; Wenzel, M.; Winter, K.; Wilde, F.; Schramm, A.; Ebeling, M.; Kasper, R.; Scheurer, M.; Sakkas, A. Comparison of Anatomical Preformed Titanium Implants and Patient-Specific CAD/CAM Implants in the Primary Reconstruction of Isolated Orbital Fractures—A Retrospective Study. *J. Pers. Med.* **2023**, *13*, 846.
9. Rodríguez-Contreras, A.; Torres, D.; Piñera-Avellaneda, D.; Pérez-Palou, L.; Ortiz-Hernández, M.; Ginebra, M.P.; Calero, J.A.; Manero, J.M.; Rupérez, E. Dual-Action Effect of Gallium and Silver Providing Osseointegration and Antibacterial Properties to Calcium Titanate Coatings on Porous Titanium Implants. *Int. J. Mol. Sci.* **2023**, *24*, 8762. [CrossRef]
10. Xing, L.; Song, H.; Wei, J.; Wang, X.; Yang, Y.; Zhe, P.; Luan, M.; Xu, J. Influence of a Composite Polylysine-Polydopamine-Quaternary Ammonium Salt Coating on Titanium on Its Ostogenic and Antibacterial Performance. *Molecules* **2023**, *28*, 4120. [CrossRef]
11. Zhang, Q.; Chen, Y.; Li, B.; Wang, C.; Wu, L.; Yao, J. Tribological Behavior of Ti-Coated Diamond/Copper Composite Coating Fabricated via Supersonic Laser Deposition. *Lubricants* **2023**, *11*, 216. [CrossRef]
12. Zuo, S.; Wang, L.; Zhang, J.; Yang, G.; Xu, Y.; Hu, Y.; Cheng, F.; Hu, X. Comparative Study of Resin and Silane Coupling Agents Coating Treatments on Bonding Strength Improvement of Titanium and Carbon Fiber Composites. *Coatings* **2023**, *13*, 903. [CrossRef]
13. Aljaafari, A.; Ahmed, F.; Shaalan, N.M.; Kumar, S.; Alsulami, A. Enhanced Thermal Stability of Sputtered TiN Thin Films for Their Applications as Diffusion Barriers against Copper Interconnect. *Inorganics* **2023**, *11*, 204. [CrossRef]
14. Cui, C.; Yang, C. Mechanical Properties and Wear Resistance of CrSiN Coating Fabricated by Magnetron Sputtering on W18Cr4V Steel. *Coatings* **2023**, *13*, 889. [CrossRef]
15. Eleutério, T.; Sério, S.; Vasconcelos, H.C. Growth of Nanostructured TiO2 Thin Films onto Lignocellulosic Fibers through Reactive DC Magnetron Sputtering: A XRD and SEM Study. *Coatings* **2023**, *13*, 922. [CrossRef]
16. Ju, Y.; Ai, L.; Qi, X.; Li, J.; Song, W. Review on Hydrophobic Thin Films Prepared Using Magnetron Sputtering Deposition. *Materials* **2023**, *16*, 3764. [CrossRef] [PubMed]
17. Kotliarenko, A.; Azzolini, O.; Cisternino, S.; El Idrissi, M.; Esposito, J.; Keppel, G.; Pira, C.; Taibi, A. First Results on Zinc Oxide Thick Film Deposition by Inverted Magnetron Sputtering for Cyclotron Solid Targets Production. *Materials* **2023**, *16*, 3810. [CrossRef] [PubMed]
18. Pang, Y.; Zhao, N.; Ruan, Y.; Sun, L.; Zhang, C. Effects of Oxygen Partial Pressure and Thermal Annealing on the Electrical Properties and High-Temperature Stability of Pt Thin-Film Resistors. *Chemosensors* **2023**, *11*, 285. [CrossRef]
19. Pokora, P.; Wojcieszak, D.; Mazur, P.; Kalisz, M.; Sikora, M. Influence of Co-Content on the Optical and Structural Properties of TiOx Thin Films Prepared by Gas Impulse Magnetron Sputtering. *Coatings* **2023**, *13*, 955. [CrossRef]
20. Rocha-Cuervo, J.J.; Uribe-Lam, E.; Treviño-Quintanilla, C.D.; Melo-Maximo, D.V. Sputtering Plasma Effect on Zinc Oxide Thin Films Produced on Photopolymer Substrates. *Polymers* **2023**, *15*, 2283. [CrossRef]
21. Sasani Ghamsari, M. Development of Thin Film Fabrication Using Magnetron Sputtering. *Metals* **2023**, *13*, 963. [CrossRef]
22. Bochenek, D.; Chrobak, A.; Ziółkowski, G. Electric and Magnetic Properties of the Multiferroic Composites Made Based on Pb(Fe1/2Nb1/2)1−xMnxO3 and the Nickel-Zinc Ferrite. *Materials* **2023**, *16*, 3785. [PubMed]
23. Mi, D.; Zhao, Z.; Bai, H. Effects of Orientation and Dispersion on Electrical Conductivity and Mechanical Properties of Carbon Nanotube/Polypropylene Composite. *Polymers* **2023**, *15*, 2370. [CrossRef] [PubMed]
24. Mollamohammadi, F.; Faridnouri, H.; Zare, E.N. Electrochemical Biosensing of L-DOPA Using Tyrosinase Immobilized on Carboxymethyl Starch-Graft-Polyaniline@MWCNTs Nanocomposite. *Biosensors* **2023**, *13*, 562. [CrossRef] [PubMed]
25. Nguyen, Q.T.; Vu, D.L.; Le, C.D.; Ahn, K.K. Enhancing the Performance of Triboelectric Generator: A Novel Approach Using Solid–Liquid Interface-Treated Foam and Metal Contacts. *Polymers* **2023**, *15*, 2392. [PubMed]
26. Regis, V.; Šadl, M.; Brennecka, G.; Bradeško, A.; Tomc, U.; Uršič, H. Investigation of Structural and Electrical Properties of Al2O3/Al Composites Prepared by Aerosol Co-Deposition. *Crystals* **2023**, *13*, 850. [CrossRef]
27. Stefano, M.; Cordella, F.; Cozza, G.; Ciancio, A.L.; Zollo, L. Study of Electrical Neural Stimulation Effects Using Extraneural and Intraneural Electrodes. *Appl. Sci.* **2023**, *13*, 6267. [CrossRef]
28. Thangamani, G.; Thangaraj, M.; Anand, P.I.; Mani, J.; Karkalos, N.E.; Papazoglou, E.L.; Karmiris-Obratański, P. Influence of Silver-Coated Tool Electrode on Electrochemical Micromachining of Incoloy 825. *Coatings* **2023**, *13*, 963. [CrossRef]
29. Lu, Q.; Han, Z.; Hu, L.; Tian, F. An Infrared and Visible Image Fusion Algorithm Method Based on a Dual Bilateral Least Squares Hybrid Filter. *Electronics* **2023**, *12*, 2292. [CrossRef]
30. Lv, J.; Hui, T.; Zhi, Y.; Xu, Y. Infrared Image Caption Based on Object-Oriented Attention. *Entropy* **2023**, *25*, 826. [CrossRef]
31. Xu, X.; Shen, Y.; Han, S. Dense-FG: A Fusion GAN Model by Using Densely Connected Blocks to Fuse Infrared and Visible Images. *Appl. Sci.* **2023**, *13*, 4684. [CrossRef]

32. Zhao, L.; Jiao, J.; Yang, L.; Pan, W.; Zeng, F.; Li, X.; Chen, F. A CNN-Based Layer-Adaptive GCPs Extraction Method for TIR Remote Sensing Images. *Remote Sens.* **2023**, *15*, 2628. [CrossRef]
33. Han, H.R. Stealth, electromagnetic interception, and electrical properties of aluminum sputtered clothing materials—Focusing on the density change. *Costume Cult. Res.* **2022**, *30*, 579. [CrossRef]

MDPI
St. Alban-Anlage 66
4052 Basel
Switzerland
www.mdpi.com

Coatings Editorial Office
E-mail: coatings@mdpi.com
www.mdpi.com/journal/coatings

www.ingramcontent.com/pod-product-compliance
Lightning Source LLC
Chambersburg PA
CBHW061630210926
51850CB00009B/2501
9783725801930